Lecture Notes in Ma

A collection of informal reports and seminars
Edited by A. Dold, Heidelberg and B. Eckmann, Zürich

Series: Mathematical Systems Theory · Advisers: R. E. Kalman and G. P. Szegö

60

Seminar on Differential Equations and Dynamical Systems

Edited by G. Stephen Jones
Institute for Fluid Dynamics and Applied Mathematics
University of Maryland, College Park, Maryland

Seminar Lectures at the University of Maryland in August 1967
given by A. Cellina, G. M. Dunkel,
G. S. Jones, J. Kato, A. Strauss, J. A. Yorke, T. Yoshizawa

1968

Springer-Verlag Berlin · Heidelberg · New York

Library of Congress Catalog Card Number 68-31622. Printed in Germany. Title No. 3666

PREFACE

It can be well argued that the informal working seminar is the most valuable vehicle for the bringing together of ideas in a lively research effort. Researchers and students alike profit by seeing results presented in preliminary form before most of the original motivating factors are polished away. This collection of papers was presented in such a seminar at the University of Maryland and it has been the intent to preserve the informality of original presentation as fully as possible.

Recent results in stability, control, and hereditary dependence for differential equations and geometric invariance in dynamical systems is the general theme of this collection. Many results are only sketchily presented and can be expected to be eventually published in fuller detail elsewhere. It is felt that this material can be a valuable supplement to standard texts for use in advanced topic courses in the theory of differential equations.

The editor and authors of this work would like to thank the Institute for Fluid Dynamics and Applied Mathematics of the University of Maryland for sponsoring the seminar which has resulted in this report. We are also indebted to the National Science Foundation for financial support. For an excellent job in typing this text we are indebted to Mrs. Madeleine Mills.

CONTENTS

LIST OF CONTRIBUTORS

Numbers in parentheses indicate the pages on which the authors' contributions begin.

ARRIGO CELLINA, Department of Mathematics, University of Maryland, College Park, Maryland (37). Partially supported by NSF Grant GP-6167 and by the CNR, Comitato per la Matematica, Gruppo 11.

GREGORY M. DUNKEL, Institute for Fluid Dynamics and Applied Mathematics, University of Maryland, College Park, Maryland (92). Partially supported by NSF Grant GP-6114 and by NIH Training Grant DE-00182.

G. STEPHEN JONES, Institute for Fluid Dynamics and Applied Mathematics, University of Maryland, College Park, Maryland (7). Partially supported by NSF Grants GP-6114 and GP-7846.

JUNJI KATO, Mathematical Institute, Tôhoku University, Sendai, Japan (26, 55, 83). Partially supported by the Sakkokai Foundations.

AARON STRAUSS, Department of Mathematics, University of Maryland, College Park, Maryland (1, 76). Partially supported by NSF Grant GP-6167.

JAMES A. YORKE, Institute for Fluid Dynamics and Applied Mathematics, University of Maryland, College Park, Maryland (31, 48, 65, 100). Partially supported by NSF Grant GP-6114.

TARO YOSHIZAWA, Mathematical Institute, Tôhoku University, Sendai, Japan (20). Partially supported by NSF Grant GP-5998.

RECENT RESULTS IN PERTURBATION THEORY

by

Aaron Strauss

I want to present several results obtained recently by James Yorke and me concerning the preservation of stability under perturbations. Consider the ordinary differential equation

(E) $$x' = f(t,x)$$

and its perturbation

(P) $$y' = f(t,y) + g(t,y) ,$$

where f and g are continuous on $[0,\infty) \times R^n$. Assume the origin 0 is EvUAS (eventually uniform-asymptotically stable) for (E). I do not assume *a priori* that solutions of (E) or (P) are unique and I do not require that $f(t,0) = 0$ or $g(t,0) = 0$. I want to perturb (E) by a function g such that 0 is EvUAS for (P) for *every* f which belongs to some pre-assigned class. This idea is in contrast to the idea of total stability, in which the bounds on g depend on f.

Let F be some class of functions. Define

$$G = G(F) = \{g : \forall f \in F , \ 0 \text{ is EvUAS for (E)} \Rightarrow 0 \text{ is EvUAS for (P)}\} ,$$

$$\mathcal{H} = \mathcal{H}(\mathcal{G}) = \{h \in \mathcal{G} : h \text{ is independent of } x\} .$$

Then, for a given $\mathcal{F}$, I want lower bounds on $\mathcal{G}$ and $\mathcal{H}$. For the classes considered here $\mathcal{G}$ and $\mathcal{H}$ are closely related to the class of diminishing functions, whose definition will now be given.

A continuous function g is called *diminishing:* if there exists $r > 0$ such that for every $m \in (0,r)$,

$$\text{(1)} \qquad \sup \left| \int_t^{t+u} g(s,z(s))ds \right| \to 0 \quad \text{as} \quad t \to \infty ,$$

where the supremum is taken with respect to all $u \in [0,1]$ and all continuous functions $z(\cdot)$ satisfying

$$m \leq |z(s)| \leq r \quad \text{for} \quad t \leq s \leq t+1 ;$$

absolutely diminishing: if there exists $r > 0$ such that for every $m \in (0,r)$,

$$\sup \int_t^{t+1} |g(s,z(s))| ds \to 0 \quad \text{as} \quad t \to \infty ,$$

where the supremum is taken with respect to all $z(\cdot)$ as above.

Examples. The function

$$h(t) = (t \sin t^3, t \cos t^3)$$

in two dimensions is diminishing but not absolutely diminishing; indeed,

$\|h(t)\| = t$. The scalar function

$$g(t,x) = t(t^2x^2+1)^{-1}$$

on $[1,\infty)$ is absolutely diminishing even though $g(t,0) = t$.

Proposition 1 . *Let* $|g(t,x)| \leq \gamma(t)$ *for* $t \geq 0$ *and* $|x| \leq 1$. *Suppose* $\gamma(t) \to 0$ *as* $t \to \infty$ *or* γ *belongs to* L^p *on* $[0,\infty)$ *for some* $p \geq 1$, *or more generally, suppose*

$$\int_t^{t+1} \gamma(s)ds \to 0 \quad as \quad t \to \infty .$$

Then g *is absolutely diminishing.*

Proposition 2 . *A function* $h = h(t)$ *is diminishing if and only if*

$$\sup_{0\leq u\leq 1} \left| \int_t^{t+u} h(s)ds \right| \to 0 \quad as \quad t \to \infty .$$

It is absolutely diminishing if and only if

$$\int_t^{t+1} |h(s)|ds \to 0 \quad as \quad t \to \infty .$$

Theorem 1 . *Let* F *be the class of Lipschitz functions, i.e., functions* f *for which there exists* $L > 0$ *such that*

$$|f(t,x) - f(t,y)| \leqslant L|x - y|$$

for all $t \geqslant 0$, $|x| \leqslant r$, *and* $|y| \leqslant r$. *Then* G *contains all diminishing functions* $g(t,x)$ *and* H *is precisely the class of diminishing functions* $h(t)$. *Furthermore,* G *contains all Lipschitz functions* g *satisfying*

$$(2) \qquad \sup_{0 \leqslant u \leqslant 1} \left| \int_t^{t+u} g(s,x)ds \right| \to 0 \quad \textit{as} \quad t \to \infty$$

for each fixed x *satisfying* $0 < |x| \leqslant r$.

Remarks. Functions satisfying (2) might not be diminishing. The point is, by demanding that g be Lipschitz, one can use condition (2) rather than (1) to see if g belongs to G . Certainly (2) is easier to verify than (1). Furthermore, (2) implies that if the matrix $B(t)$ is diminishing and bounded and if $k(x)$ is Lipschitz, then $B(t)k(x) \in G$. (I do not assume $k(0) = 0$.) It can also be shown that $B(t)x \in G$ if $B(t)$ is diminishing and diagonal. If $B(t)$ is merely diminishing, it is not known whether or not $B(t)x$ belongs to G .

Theorem 2 . *Let* f *be periodic in* t , *i.e.,* $f(t + \omega,x) = f(t,x)$ *for all* $t \geqslant 0$, *all* $x \in R^n$, *and some* $\omega > 0$. *(I do not assume* (E) *has uniqueness.) Then* G *contains all diminishing functions* $g(t,x)$ *and* H *is again the class of diminishing functions* $h(t)$.

Theorem 3 . *Let* f *satisfy the (Euclidean) inner product condition*

$$\langle f(t,x) - f(t,y),\ x - y \rangle \leq L\,||x - y||^{2} .$$

Then G *and* H *each contain all absolutely diminishing functions (but not all diminishing functions).*

Theorem 4 . *Let* $f(t,x) = A(t)x$. *Then*

$$G = \{g + \phi : g \text{ is absolutely diminishing and } \phi(t,x) = o(|x|)\}$$

and H *contains all absolutely diminishing function (but neither contains all diminishing functions).*

Remarks. In Theorems 3 and 4, in contrast to the case in Theorems 1 and 2, Yorke and I have been unable to precisely indentify H . However, we can prove in each case that H is closed under the pointwise addition of any two of its members. We conjecture that, in each of the above four theorems, G is not closed under addition.

An upper bound on the size of F needed to obtain nontrivial G and H is provided by the final result. Let F be the class of functions f which are continuous in x uniformly with respect to t for t in $[0,\infty)$ and for which all solutions of (E) are unique to the right.

Theorem 5 . *For* F *as above,* H *does not contain all the absolu*[cut off]*ly diminishing functions. In fact, given any function* $\lambda(\cdot)$ *satisfying* $\lambda(t) > 0$ *for* $t \geq 0$, H *does not contain all* C^∞ *functions satisfying*

$$|h(t)| \leq \lambda(t) \quad \textit{for} \quad t \geq 0 .$$

THE EXISTENCE OF CRITICAL POINTS IN GENERALIZED DYNAMICAL SYSTEMS

by

G. Stephen Jones

1. *Introduction*

Let X be a closed locally compact convex subset of a Banach space, let R^+ denote the nonnegative real numbers, and let I denote the invertal $[0,1]$. Let $\pi : R^+ \times X \to X$ be a continuous mapping such that $\pi(0;x) = x$ and $\pi(t_1;\pi(t_2;x)) = \pi(t_1 + t_2;x)$ for all $x \in X$ and t_1, t_2 in R^+. Considering the mapping $\pi(t;\cdot)$ as a family of mappings of X into X, we note that this family forms a semigroup. We shall, for purposes of drawing attention to the fact that any of the results presented here hold for dynamical systems on X defined in the usual way (see [1]), refer to π as a *generalized dynamical system*.

The family of solutions of autonomous ordinary differential equations as functions of their initial data and parameterized by time form the most classical examples of dynamical systems. When autonomous hereditary differential equations are considered in the same way, the result is a generalized dynamical system such as defined where the base space X may of necessity be quite general (see [2]) . This is particulatly true if the hereditary arguments are state dependent. Certain classes of partial differential equations and integral equations may also be productively studied by considering their families of solutions as generalized dynamical systems.

Let S be a set in X. If $\pi : R^+ \times X \to X$ is a generalized dynamical system and for some $t > 0$ we have that $\pi(t;\bar{S}) \subset S$ then S is said to be *constrained* under the action of π. If for some $t > 0$ $\pi(t;\bar{S})$ is contained in a compact subset of S, then S is said to be *compactly constrained* by π at t. If for each $x \in \bar{S}$ there exists some $t = t(x) > 0$ such that $\pi(t;x) \in S$, then S is said to be *weakly constrained* by π. By $\pi(\infty;x)$ we denote the set of limit points of all sequences $\{\pi(t_n;x)\}$ where $\{t_n\} \subset R^+$ is any sequence tending to ∞. For $S \subset X$, $\pi(\infty;S) = \bigcup\{\pi(\infty;x) : x \in S\}$. It is also convenient to define ϕ to be a function $\phi : I \times R^+ \to R^+$ which is positive except possibly on $I \times \{0\}$ and strictly increasing in its second argument.

A problem of some importance in the study of generalized dynamical systems is the determination of the presence of equilibrium states or critical points; that is, determining the presence of points $x^* \in X$ such that $\pi(t;x^*) = x^*$ for all $t \in R^+$. It is to this problem that this paper is addressed. The results presented are set forth in the following three theorems:

Theorem 1. *Let* S *be an open convex set in* X *and let* $\pi : R^+ \times X \to X$ *be a generalized dynamical system. If* π *compactly constrains* S, *then* π *has a critical point.*

Theorem 2. *Let* S *be an open convex set in* X *and let* $\pi : R^+ \times X \to X$ *be a generalized dynamical system. Let* $\pi(\infty;\bar{S})$ *be a compact subset of* S, *let* r *be a retraction of* X *onto the closed convex hull of* $\pi(\infty;\bar{S})$ *, and for each* $\lambda \in I$ *let*

$$\pi_\lambda(t;x) = (1 - \lambda)\ \pi(t;x) + \lambda r\pi(t;x) .$$

If for all t_1 *and* t_2 *in* R^+*,*

$$\pi_\lambda(t_1;\pi_\lambda(t_2;\bar{S})) \subset \pi_\lambda(t_1 + \phi(\lambda,t_2);\ \bar{S}) ,$$

then π *has a critical point in* S *.*

Theorem 3 *. Let* S *be an open convex set in* X *and let* $\pi : R^+ \times X \to X$ *be a generalized dynamical system. Suppose there exists a continuous function* $\zeta : I \times R^+ \times X \to X$ *such that for all* $\lambda \in I$ *and all* t_1 *and* t_2 *in* R^+*,*

$$\zeta(\lambda;t_1;\zeta(\lambda;t_2;\bar{S})) \subset \zeta(\lambda;t_1 + \phi(\lambda,t_2);\ \bar{S})$$

and $\zeta(0;\cdot;\cdot) = \pi$ *. If* $\zeta(\lambda;\cdot;\cdot)$ *for each* $\lambda \in I$ *weakly constrains* S *,* $\zeta(1;\cdot;\cdot)$ *compactly constrains* S *, and* $\bigcup\{\zeta(\lambda;\infty;\ \bar{S}) : \lambda \in I\}$ *is compact, then* π *has a critical point.*

Clearly Theorem 1 is a special case of both Theorem 2 and Theorem 3.

2. *Comments.*

We note that, although X has been specified to lie in a Banach space , all results stated and supporting proofs presented are valid for X taken in a metrizable complete locally convex linear space. This generalization is important, for instance, in considering functional equations such as presented in [2].

Let a set S be called homeomorphically convex if there exists a homeomorphism $\theta : X \to X$ which sends S onto a convex set. We note that the convexity condition on S in Theorems 1, 2, and 3 can be weakened to homeomorphic convexity.

A set S is said to be *externally convex* if the union of S with all of the bounded components of its complement is convex. If this union is homeomorphically convex, then S is said to be *homeomorphically externally convex*. If the generalized dynamical system π specified in our theorems is such that $x \neq y$ implies $\pi(t;x) \neq \pi(t;y)$ for all x and y in X and $t \in R^+$, then the condition that S be convex in our theorems can be weakened to the requirement that S be externally homeomorphically convex. In particular, this is true if π is a dynamical system as defined in [1].

3. *Preliminary Lemmas*

As a convenience in the proofs of the theorems stated we shall in this section present and prove several lemmas.

Lemma 1 . *Let* S *be an open set in* X *and let* $\pi : R^+ \times X \to X$ *be a generalized dynamical system which compactly constrains* S *at* t_o . *Then there exists* $t_1 > 0$ *such that* $\pi(t;\bar{S}) \subset S$ *for all* $t \geq t_1$.

Proof. By hypothesis there exists a compact set $U_o \subset S$ and $\delta_o > 0$ such that $\pi(t;\bar{S}) \subset U_o$ for all $t \in [t_1, t_1 + 2\delta_o]$. Furthermore, we can choose a compact set $U_1 \subset S$ and $\delta_1 > 0$ such that $\pi(t;U_o) \subset U_1$ for all $t \in [0,\delta_1]$. Letting $h = \min\{\delta_o, \delta_1\}$ there exist a positive integer n such that nh and $(n + 1)h$ are contained in $[t_1, t_1 + 2\delta_o]$, so $\pi(nh;\bar{S}) \cup \pi((n + 1)h,\bar{S}) \subset U_o$. But this implies that for any pair of positive integers k_o and k_1 such that $k_o + k_1 > 0$ that $\pi(k_o nh,\bar{S}) \cup \pi(k_1(n + 1)h,\bar{S}) \subset U_o$ and in fact that for every number of the form $(k_o n + k_1(n + 1))h$ that $\pi((k_o n + k_1(n + 1))h;\bar{S}) \subset U_o$. But every integer $m \geq (n - 1)n$ can be expressed in the form $k_o n + k_1(n + 1)$ so $\pi(mh;\bar{S}) \subset U_o$ for all $m \geq (n - 1)n$. Hence for all $t \geq (n - 1)nh$ we have that $\pi(t;\bar{S}) \subset U_1 \subset S$ and our lemma is proved.

Our next lemma is a simple variation of the Homotopy type argument for the existence of fixed points under parametric variation.

Lemma 2 . *Let* S *be an open convex subset of* X , *let* $f : I \times X \to X$ *be a continuous function, and let* $f(0,\cdot)$ *have a fixed point in* S . *Then either* $f(1;\cdot)$ *has a fixed point in* S *or for every compact subset* U *of* S *there exists* $\lambda \in I$ *such that* $f(\lambda;\cdot)$ *has a fixed point in* $S \setminus U$.

Proof. Suppose there exists a compact set $U \subset S$ such that for all $\lambda \in [0,1]$, $f(\lambda;\cdot)$ does not have a fixed point in $S \setminus U$. Then there exists a compact convex neighborhood V of U contained in S such that $f(\lambda;\cdot) \mid V$ has no fixed point on the boundary of V for all $\lambda \in I$. Hence by the standard homotopy argument (See [4]) $f(1,\cdot)$ must have a fixed point in V and our lemma is proved.

Lemma 3. *Let* S *be an open convex set contained in* X, *let* $\psi : R^+ \to R^+$ *be a positive strictly increasing function on* $R^+ \setminus \{0\}$ *and let* $q : R^+ \times X \to X$ *be a continuous map such that* $q(0,x) = x$ *and for all* t_1 *and* t_2 *in* R^+, $q(t_1;q(t_2;\bar{S})) \subset q(t_1 + \psi(t_2);\bar{S})$. *If* q *constrains* S *compactly for all* t *sufficiently large, then* $q(t;\cdot)$ *has a fixed point for every* $t \in R^+$.

Proof. By hypothesis q compactly constrains S for all t_o sufficiently large so there exists $t^* > 0$ such that for all $t \geq t^*$, $q(t,\bar{S})$ is contained in a compact subset U of S. From the Tychonov fixed point theorem (see [5]) we may conclude that for all $t \geq t^*$ there exists $x_t \in S$ such that $q(t;x_t) = x_t$.

Let us suppose that $q(t;\cdot)$ does not have a fixed point for every $t \in (0,t^*)$. Then by Lemma 2 there exists $t_1 \in (0,t^*)$ and $x_{t_1} \in \bar{S} \setminus U$ such that $q(t_1;x_{t_1}) = x_{t_1}$. Now $x_{t_1} \in \bar{S} \setminus U$ implies $x_{t_1} = q^n(t_1;x_{t_1}) \in \bar{S} \setminus U$ for all integer $n \geq 1$ and $q^n(t_1;x_{t_1})$ by hypothesis is contained in $q(t_1 + (n-1)\psi(t_1);\bar{S})$. But $q(t_1 + (n-1)\psi(t_1);\bar{S})$ is contained in U for all n sufficiently large so we can only conclude that $x_{t_1} \in U$. By contradiction, therefore, it follows that $q(t,\cdot)$ must have a fixed point in S for all t in $(0,t^*)$ and hence for all $t \in R^+$. Thus Lemma 3 is proved.

Lemma 4 . *Let* S *be an open set contained in* X $\psi : R^+ \to R^+$ *be a positive strictly increasing function on* $R^+ \setminus \{0\}$ *and let* $q : R^+ \times X \to X$ *be a continuous map such that* $q(0,x) = x$ *and for all* t_1 *and* t_2 *in* R^+ , $q(t_1;q(t_2;\bar{S}) \subset q(t_1 + \psi(t_2);\bar{S})$. *Furthermore, let* J *be a closed interval* $[0,b]$, $b > 0$, *and let* $q(t_1;\cdot)$ *have a fixed point for each* $t \in J$. *If for all* t_1 *and* t_2 *in* J *such that* $t_1 + t_2 \in J$ *we have that* $q(t_1;q(t_2;x)) = q(t_1 + t_2;x)$ *for all* x *in* $\bar{S}$, *then there exists* x^* *in* S *such that* $q(t;x^*) = x^*$ *for all* $t \in J$.

Proof. By hypothesis $q(t;\cdot)$ has a fixed point for every $t \in J$. Let $\{t_k\} \to 0$ be a sequence in J .
Let $\{x_{t_k}\}$ denote a corresponding sequence of points in S such that $q(t_k;x_{t_k}) = x_{t_k}$. Since S is compactly constrained by q it follows that $\{x_{t_k}\}$ must be contained in a compact subset of S and therefore has a limit point x^* in S . We shall show that x^* has the property that $q(t;x^*) = x^*$ for all $t \in J$.

If $q(t;x^*)$ is not identically equal to x^* on J then we can choose τ in J such that $q(\tau;x^*) \neq x^*$. We can thus choose a closed neighborhood $N(q(\tau;x^*))$ of $q(\tau;x^*)$ not containing x^*. By continuity we can choose a neighborhood $N_o(x^*)$ of x^* with $N_o(x^*) \cap N(q(\tau;x^*))$ empty and $\varepsilon > 0$ such that $x \in N_o(x^*)$ and $|t - \tau| < \varepsilon$ imply $q(\tau;x) \in N(q(\tau;x^*))$. We can choose $t_k \in \{t_k\}$ such that $x_{t_k} \in N_o(x^*)$ and for some positive integer n, $|n(t_k) - \tau| < \varepsilon$. But $x_{t_k} = q(n(t_k); x_{t_k})$ and we have that $q(a + n(t_k - a);x_{t_k}) \in N(q(\tau,x^*))$. Since $N_o(x^*) \cap N(q(\tau;x^*))$ is empty we clearly have a contradiction. It follows that $q(t;x^*) = x^*$ for all t in J and our proof is complete.

4. *Proofs of Theorems* 1, 2, *and* 3 .

With the lemmas of the last section the proofs of Theorem 1 and Theorem 2 follow with little difficulty.

Theorem 1 . *Let* S *be an open convex set in* X *and let* $\pi : R^+ \times X \to X$ *be a generalized dynamical system. If* π *compactly constrains* S, *then* π *has a critical point.*

Proof. From lemma 3 it follows immediately that $\pi(t;\cdot)$ has a fixed point for all t in R^+. Hence by Lemma 4 with $J = R^+$ it follows that there exists x^* such that $q(t;x^*) = x^*$ for all $t \in R^+$.

Theorem 2 . *Let* S *be an open convex set in* X *and let* $\pi : R^+ \times X \to X$ *be a generalized dynamical system. Let* $\pi(\infty;\bar{S})$ *be a compact subset of* S, *let* r *be a retraction of* X *onto the closed convex hull of* $\pi(\infty;\bar{S})$, *and for each* $\lambda \in I$ *let*

$$\pi_\lambda(t;x) = (1 - \lambda)\ \pi(t;x) + \lambda r\pi(t;x) .$$

for all t_1 *and* t_2 *sufficiently large,*

$$\pi_\lambda(t_1;\pi_\lambda(t_2;\bar{S})) \subset \pi_\lambda(t_1 + \phi(\lambda,t_2);\bar{S}) ,$$

hen π *has a critical point in* S .

Proof. Let $\{\tau_k\} \to \infty$ be a sequence in R^+ and for each τ_k et $q_k : R^+ \times X \to X$ be defined by the formula

$$q_k(t;x) = \pi(t;x) , \quad \text{for} \quad t \in [0,\tau_k] ,$$

$$q_k(t;x) = \pi_{t-\tau_k}(t;x), \text{for} \quad t \in [\tau_k,\tau_k + 1] ,$$

$$q_k(t;x) = r\pi(t;x) , \quad \text{for} \quad t \geq \tau_k + 1 .$$

learly q_k for each k is continuous so by the Tychonov Fixed Point heorem q_k has a fixed point for each $t \geq \tau_k + 1$.

Suppose q_k does not have a fixed point for each $t \in [0,\tau_k + 1]$. hen by Lemma 2 there exist $t_1 \in [0,\pi_k + 1]$ and $x_{t_1} \in \bar{S} \setminus H(\pi(\infty;\bar{S}))$, here $H(\pi(\infty;\bar{S}))$ denotes the closed convex hull of $\pi(\infty;\bar{S})$, such that $(t_1,x_{t_1}) = x_{t_1}$. If $t_1 \leq \tau_k$ then clearly $x_{t_1} = \pi(nt_1;x_{t_1})$ for all ≥ 1 which, of course, implies $x_{t_1} \in \pi(\infty;\bar{S})$. Hence t_1 must be in $\tau_k,\tau_k + 1]$. But by hypothesis with t_o sufficiently large this would imply that

$$x_{t_1} = q_k^n(t_1;x_{t_1}) \in q_k(t_1 + (n-1)\phi(t_1 - \tau_k,t_1);\bar{S})$$

for all n and therefore x_{t_1} would have to lie in $H(\pi(\infty:\bar{S}))$. Hence no such t_1 can exist and we conclude that each $q_k(t;\cdot)$ must have a fixed point for each $t \in [0,\infty)$. By Lemma 4 it follows that there exists $x_k^* \in S$ such that

$$q_k(t;x_k^*) = \pi(t;x_k^*) = x_k^* \quad \text{for all} \quad t \in [0,\tau_k] .$$

It follows easily that every point x_t^* must be a critical point for π and our theorem is proved.

Theorem 3 . *Let* S *be an open convex set in* X *and let* $\pi : R^+ \times X \to X$ *be a generalized dynamical system. Suppose there exists a continuous function* $\zeta : I \times R^+ \times X \to X$ *such that for all* $\lambda \in I$ *and all* t_1 *and* t_2 *in* R^+ ,

$$\zeta(\lambda;t_1;\zeta(\lambda;t_2;\bar{S})) \subset \zeta(\lambda;t_1 + \phi(\lambda,t_2);\bar{S})$$

and $\zeta(0;\cdot;\cdot) = \pi$. *If* $\zeta(\lambda;\cdot;\cdot)$ *for each* $\lambda \in I$ *weakly constrains* $S, \zeta(1;\cdot;\cdot)$ *compactly constrains* S *for all* t_o *sufficiently large, and* $\bigcap\{\zeta(\lambda;\infty;\bar{S}) : \lambda \in I\}$ *is compact, then* π *has a critical point.*

Proof. By Lemma 3 for every $t \in R^+$ there exists x_t such that $\zeta(1;t;x_t) = x_t$. Let $U = \bigcup\{\zeta(\lambda;\infty;\bar{S}) : \lambda \in I\}$ and suppose for some $t_1 \in R^+$, $\pi(t_1;\cdot) = \zeta(0;t_1;\cdot)$ does not have a fixed point. Then by Lemma 2 for some $\lambda \in I$ there exists $x_{\lambda,t_1} \in S/U$ such that $\zeta(\lambda;t_1;x_{\lambda,t_1}) = x_{\lambda,t_1}$. But this implies $x_{\lambda,t_1} \notin \zeta(\lambda;\infty;\bar{S})$ which is, of course, absurd. Hence $\pi(t;\cdot)$ has a fixed point for all $t \in R^+$. By Lemma 4 this implies the existence of $x^* \in S$ such that $\pi(t;x^*) = x^*$ for all t and our theorem is proved.

5. *Further Remarks.*

Let S be defined as in Theorem 1 and let us suppose that $\pi : R^+ \times X \to X$ is such that $\pi(t,\bar{S})$ is compact for t sufficiently large. Let x_o be a point in S, let $B \subset S$ be a bounded convex neighborhood of x_o and let ∂B denote the boundary of B. Instead of assumming that S is constrained let us require that no closed subset of ∂B be constrained and that for all $s > 0$,

$$(1 + s)(x - x_o) \neq \pi(t;x - x_o) - \pi(t;x_o - x)$$

for all $x \in \partial B$ and all t sufficiently large. It can be shown that we can still conclude the existence of a critical point. This extension, among other possibilities allows us to establish the existence of critical points in sets which are unstable or repulsive in a certain sense. In particular, if there exists $t_1 > 0$ such that for all $t \geq t_1$ and $\lambda \in I$,

$$(\lambda - 1)\ \pi(t_1;\partial B) + \lambda\pi(t_1;\partial B) \cap B$$

is empty we have the existence of at least one critical point.

In this paper we have only considered generalized dynamical system over R^+. If one is interested in considering the existence of periodic solutions of functional equations, dynamical systems over the nonnegative integers are interesting and a theory similar to what has been developed herein is possible. The interested reader is referred to [6] and [7].

REFERENCES

[1] V. V. NEMYTSKII and V.V. STEPANOV, Qualitative theory of differential equations, Princeton University Press, (1960).

[2] G.S. JONES, Hereditary Structure in Differential Equations, Math. Systems Theory, Vol. 2 (1968).

[3] A. GRANAS, The theory of compact vector fields and some of its applications to topology of functional spaces, Rozprawy Matematyczne, polska Akademia Nauk. Instytut Matematyezmy, Warsaw (1962).

[4] J. DUGUNDJI, An extension of Tietze's theorem, Pac. J. Math. 1, 353-367 (1951).

[5] A. TYCHONOV, Ein Fixpunktstaz, Math. Ann, Vol. 111, 767-776 (1935).

[6] G.S. JONES, Asymptotic fixed point theory, Proc. Symp. Infinite Dimensional Topology, Louisiana State Univ. Press (1967).

[7] G.S. JONES, Fixed point theorems for asymptotically compact mappings, U. of Maryland Report BN-503 (1967).

STABILITY AND EXISTENCE OF PERIODIC AND ALMOST PERIODIC SOLUTIONS

by

Taro Yoshizawa

In 1950, Massera [1] discussed the existence theorems for a periodic solution of the system of differential equations

$$(1) \qquad \frac{dx}{dt} = F(t,x) \ ,$$

where x and F are n-vectors, $F(t,x)$ is defined and continuous on $I \times R^n$ (I: the interval $0 \leqq t < \infty$, R^n : n-dimensional Euclidean space), $F(t,x)$ is periodic in t of period ω and for the initial value problem, the solution of (1) is unique, and he showed that if $F(t,x)$ is linear in x or if $n = 1$, the existence of a bounded solution of (1) implies the existence of a periodic solution of period ω. Massera pointed out that for $n = 2$, the existence of a bounded solution does not necessarily imply the existence of a periodic solution of period ω and then proved that for $n = 2$, if every solution of (1) exists in the future and one of them is bounded, then there exists a periodic solution of period ω.

If $n > 2$, we need further additional conditions to prove the existence of a periodic solution. Thus, there arises the question of what kin of additional condition is required. Many authors assumed some kind of stability property of a bounded solution.

Recently, Sell [2] has shown that if a bounded solution $\phi(t)$ of

(1) is uniform-stable and if there is a $\delta_0 > 0$ such that $t_0 \in I$ and $\|x_0 - \phi(t_0)\| < \delta_0$ imply that $\|x(t;x_0,t_0) - \phi(t)\| \to 0$ as $t \to \infty$, then there exists a periodic solution of period $k\omega$ for some integer $k \geq 1$, where $x(t;x_0,t_0)$ is the solution through (t_0,x_0).

Actually, Sell's condition implies that the bounded solution $\phi(t)$ is uniform-asymptotically stable (see [3]). For an almost periodic system with zero a solution, Seifert [4] has shown that if the zero solution satisfies Sell's conditions, then the zero solution is uniform-asymptotically stable. However, for any bounded solution of an almost periodic system, the question is still open.

Under the assumptions of Sell, we cannot prove the existence of a periodic solution of period ω. Yorke and, independently, Seifert [4] gave an example which has no periodic solution of period ω, but has a uniform-asymptotically stable periodic solution of period 2ω. Therefore, if we want to have a periodic solution of period ω, we need some additional conditions. The following is a sufficient condition.

Theorem 1 . *Suppose that the periodic system* (1) *has a bounded solution* $\phi(t)$ *which is uniform-asymptotically stable and for which* $\|\phi(t)\| \leq B$ *for all* $t \geq 0$. *Letting* $S = \{x; \|x\| \leq B\}$, *if every solution starting from* S *at* $t = 0$ *is stable, then there exists a periodic solution* p(t) *of period* ω *which is uniform-asymptotically stable and satisfies* $\|p(t)\| \leq B$.

Moreover, Sell has shown in [2] that if a bounded solution $\phi(t)$ of (1) is uniform-stable and if for every $t_0 \in I$ and every $x_0 \in R^n$, we have $\|x(t;x_0,t_0) - \phi(t)\| \to 0$ as $t \to \infty$, then there exists a periodic solution of period $k\omega$ for some integer $k \geq 1$.

For the periodic system (1), Sell's condition is equivalent to saying that a bounded solution $\phi(t)$ is uniform-asymptotically stable in the large. Therefore, we can conclude that there exists a periodic solution of period ω which is uniform-asymptotically stable in the large (see [3]).

For an almost periodic system, Seifert [4] has shown that Sell's condition does not necessarily imply the globally uniform-asymptotic stability of $\phi(t)$.

Thus, we have the following theorem.

Theorem 2 . *For the periodic system* (1) , *suppose that there exists a bounded solution which is uniform-asymptotically stable in the large. Then, there exists a unique periodic solution* $p(t)$ *of* (1) *of period* ω *which is uniform-asymptotically stable in the large.*

For an almost periodic system, we can also obtain a similar theorem to Theorem 2. However, it is not easy to see the existence of a bounded solution $\phi(t)$ which is uniform-asymptotically stable in the large. We want to characterize it by the existence of Liapunov functions.

We consider the periodic system (1) and assume that $F(t,x)$ satisfies locally a Lipschitz condition with respect to x . Then, there exists an $L(\alpha) > 0$ such that

$$(2) \qquad \|F(t,x) - F(t,y)\| \leq L(\alpha)\, \|x - y\| \quad \text{if} \quad \|x\| \leq \alpha \quad \text{and} \quad \|y\| \leq \alpha \ .$$

Theorem 3 . *Assume that* $F(t,x)$ *in* (1) *is defined and continuous on* $(-\infty,\infty) \times R^n$, *is periodic in* t *of period* ω *and satisfies the condition* (2). *In order that there exists a periodic solution* $p(t)$ *of period* ω *of* (1) *such that* $\|p(t)\| < B$ *which is uniform-asymptotically stable in the large, it is necessary and sufficient that the solutions of* (1) *are uniform-bounded and uniform-ultimately bounded for bound* B *and there exists a continuous Liapunov function* $V(t,x,y)$ *defined on* $0 \leqq t < \infty$, $\|x\| \leqq B$, $\|y\| \leqq B$ *which satisfies the following conditions:*

(i) $a(\|x - y\|) \leqq V(t,x,y) \leqq b(\|x - y\|)$, *where* $a(r)$ *and* $b(r)$ *are continuous increasing positive definite function for* $0 \leqq r \leqq 2B$,

(ii) $|V(t,x_1,y_1) - V(t,x_2,y_2)| \leqq K\{\|x_1 - x_2\| + \|y_1 - y_2\|\}$, *where* $K > 0$ *is a constant,*

(iii) *in the domain* $0 \leqq t < \infty$, $\|x\| < B$, $\|y\| < B$, *we have*

$$(3) \qquad \dot{V}(t,x,y) = \overline{\lim_{h \to 0+}} \frac{1}{h}\{V(t + h, x + hF(t,x), y + hF(t,y)) - V(t,x,y)\}$$

$$\leqq -cV(t,x,y) ,$$

where $c > 0$ *is a constant.*

Next, we shall consider an almost periodic system

$$(4) \qquad \frac{dx}{dt} = G(t,x) ,$$

where $G(t,x)$ is defined and continuous on $(-\infty,\infty) \times R^n$, is almost periodic in t uniformly with respect to $x \in S$ for any compact set S in R^n and satisfies

(5) $\|G(t,x) - G(t,y)\| \leq L(\alpha)\,\|x - y\|$ if $\|x\| \leq \alpha$ and $\|y\| \leq \alpha$.

Theorem 4 . *In order that there exists an almost periodic solution of* (4) *which is uniform-asymptotically stable in the large, it is necessary and sufficient that there exists a positive constant* B *such that the solutions of* (4) *are uniform-bounded and uniform-ultimately bounded for bound* B *and that there exists a continuous Liapunov function* $V(t,x,y)$ *defined on* $0 \leq t < \infty$, $\|x\| < B + \varepsilon_0$, $\|y\| < B + \varepsilon_0$ *which satisfies the conditions* (i) , (ii) *and* (iii) *in Theorem* 3 *for* $B + \varepsilon_0$ *instead of* B , *where* $\varepsilon_0 > 0$ *is arbitrarily small but fixed.*

Remark 1. A necessary and sufficient condition in order that the solutions are uniform-bounded and uniform-ultimately bounded is given by the existence of some Liapunov function.

Remark 2 . In Theorems 3 and 4 , the condition (3) can be replaced by

$$\dot{V}(t,x,y) \leq -c(\|x - y\|) ,$$

where $c(r)$ is continuous and positive definite.

For the details, see [5].

REFERENCES

[1] J. L. MASSERA, The existence of periodic solutions of systems of differential equations, Duke Math. J. 17(1950), 457-475.

[2] G. R. SELL, Periodic solutions and asymptotic stability, J. Diff. Eqs., 2(1966), 143-157.

[3] T. YOSHIZAWA, Stability and existence of a periodic solution, J. Diff. Eqs. (to appear).

[4] G. SEIFERT, Almost periodic solutions and asymptotic stability, J. Math. Anal. Appl., (to appear).

[5] T. YOSHIZAWA, Existence of a globally uniform-asymptotically stable periodic and almost periodic solution, Tohoku Math. J. (to appear).

ASYMPTOTIC EQUIVALENCE

by

Junji Kato

For a linear system and its perturbed system many authors have discussed the problem of asymptotic equivalence. The problem of the existence of a solution which tends to zero as $t \to \infty$, or in short an 0-curve, is closely connected with this problem.

Hartman and Onuchic [1] obtained a good result about the existence of an 0-curve of a system

$$\dot{x} = Ax + f(t,x)$$

by using results due to Massera and Schäffer [3]. Not knowing their result, the author obtained the same result by utilizing a kind of boundary value problem [2]. He has also considered the same problem for functional differential equations and obtained the following result [2]. Here, C denotes a space of continuous R^n-valued functions defined on $[-r,0]$ for a given constant $r > 0$. For a continuous R^n-valued function $x(s)$, x_t denotes a function in C such that

$$x_t(\theta) = x(t + \theta) \quad \text{for} \quad \theta \in [-r,0] .$$

Let $||\phi||_r$ be a norm in C defined by

$$||\phi||_r = \sup\{\, ||\phi(\theta)|| \;\; ; \;\; \theta \in [-r,0]\} \quad .$$

Theorem 1 . *Consider a system of functional differential equations*

(1) $$\dot{x}(t) = F(x_t) + X(t,x_t) \ ,$$

and assume that $F(\phi)$, $X(t,\phi)$ *are continuous on* $[0,\infty) \times C$ *and* $F(\phi)$ *is linear in* ϕ . *Let* p *be the maximum of the multiplicites of the roots with zero real parts of the characteristic equation with respect to the system*

$$\dot{x}(t) = F(x_t)$$

if there exist roots with zero real parts, and otherwise set $p = 1$. *If there exists a continuous function* $\lambda(t,\alpha)$ *such that*

$$||X(t,\phi)|| \leq \lambda(t,\alpha) \quad \textit{for any} \;\; \phi \ , \quad ||\phi||_r \leq \alpha \quad ,$$

and

$$\int^{\infty} t^{p-1} \, \lambda(t,\alpha) dt < \infty \ ,$$

then there exists an 0-*curve of the system* (1) .

Now, consider the system

(2) $$\dot{x}(t) = Ax(t - r(t)) ,$$

where A is a constant matrix and $r(t)$ is a continuous function such that $0 \leqq r(t) \leqq r$ for all $t \geqq 0$. Then, applying Theorem 1, we can prove the following theorems.

Theorem 2 . *Under one of the following conditions*

(i) $$\int^{\infty} r(t)dt < \infty ;$$

(ii) $$\int^{\infty} r(t)^2 dt < \infty$$

and $r(t)$ *satisfies*

$$|r(t) - r(t')| \leqq L|t - t'|$$

for a constant $L > 0$ *and large* t, t' ;

(iii) $$\int^{\infty} r(t)^2 dt < \infty$$

and $r(t)$ *is monotone for large* t ;

for any solution $x(t)$ *of* (2) *there exists a constant vector* c *such that*

(3) $$\exp[-At + \int_0^t A^2 r(s)ds]x(t) \to c \quad \textit{as} \quad t \to \infty ,$$

and conversely, for any constant n-vector c *we can find a solution* $x(t)$ *of* (2) *satisfying* (3) .

Theorem 3 . *Let* r_0 *be a constant such that* $0 < r_0 \leqq r$, *and set* $p = \max\{\ell, 2m\}$, *where* ℓ *is the largest multiplicity of the characteristic roots of* A *and* m *is the multiplicity of the characteristic root* $-\frac{1}{r_0 e}$ *of* A , *if such a root exists.*

Suppose that $r(t)$ *satisfies*

$$\int^{\infty} t^{p-1}|r(t) - r_0|dt < \infty.$$

Then, if $e^{\rho t}c(t)$ *is a solution of the system*

$$\dot{x}(t) = Ax(t - r_0) \tag{4}$$

for a bounded function $c(t)$ *and a constant* ρ *(and hence* ρ *may be a root of the characteristic equation with respect to the system* (4)), *then there exists a solution* $x(t)$ *of* (2) *such that*

$$x(t)e^{-\rho t} - c(t) \to 0 \quad \text{as} \quad t \to \infty , \tag{5}$$

and conversely, if $x(t)$ *is a solution of* (2) *such that* $x(t)e^{-\rho t}$ *is bounded for a root* ρ *of the characteristic equation with respect to the system* (4) , *then there exists a bounded function* $c(t)$ *satisfying* (5) *such that* $e^{\rho t}c(t)$ *is a solution of* (4) .

This is contained in the paper presented to the U.S.-Japan Seminar at Minneapolis on June, 1967.

REFERENCES

[1] P. HARTMAN and N. ONUCHIC, On the asymptotic integration of ordinary differential equations, Pacific J. Math., 13(1963), 1193-1207.

[2] J. KATO, On the existence of 0-curves, Tohoku Math. J., 19(1967), 49-62.

[3] J. L. MASSERA and J. J. SCHÄFFER, Linear differential equations and functional analysis IV, Math. Ann., 139(1960), 287-342.

EXTENDING LIAPUNOV'S SECOND METHOD TO NON-LIPSCHITZ LIAPUNOV FUNCTIONS

by

James A. Yorke

The behavior of solutions of an ordinary differential equation

$$\text{(E)} \qquad \frac{d}{dt} x = f(t,x) ,$$

where $f : U \to R^n$ is continuous on the open set $U \subset R \times R^n$, is frequently studied by means of a continuous function $V : U \to R$. It is sometimes unnecessar to know the solutions explicitly. If for example V is independent of t, $V(x_o) = 0$ for some x_o, $V(x) > 0$ for $x \neq x_o$, and if for each solution ϕ of (E), $V(\phi(t))$ is a monotonically decreasing function of t for $t \geq 0$, then x_o is a stable critical point of (E). For V a C^1 function, Liapunov defined

$$\dot{V}(t,x) = \frac{\partial}{\partial t} V(t,x) + \langle \text{grad}_x V(t,x) , f(t,x) \rangle$$

where $\langle \cdot , \cdot \rangle$ denotes the inner product in R^n. He observed that for any solution ϕ, $\dot{V}(t,\phi(t)) = \frac{d}{dt} V(t,\phi(t))$;
hence the rate of change of $V(t,\phi(t))$ can be calculated directly from V and f without knowing the solutions when V is a C^1 function. Sometimes a likely function V is not C^1, and for converse theorems frequently the most difficult problem is proving V can be chosen to be a smooth function. A theory was thus developed for $V \in C^o$ (V locally Lipschitz in x) primarily by Yoshizawa [2, p. 4] with earlier results by Okamura [1]. We will mean by $\phi(\cdot;t_o,x_o)$ that ϕ is a solution of (E) such that $\phi(t_o) = x_o$. When we refer to the domain of ϕ, we will assume that ϕ cannot be extended to a larger domain and still be a solution. Define for a

solution $\phi = \phi(\cdot;t,x)$

$$(1) \qquad V'(t,x) = \liminf_{\tau \to +0} \tau^{-1}[V(t+\tau, \phi(t+\tau)) - V(t, \phi(t))]$$

$$(2) \qquad \dot{V}(t,x) = \liminf_{\tau \to +0} \tau^{-1}[V(t+\tau, x+\tau f(t,x)) - V(t,x)]$$

In (1) and (2) we use the so-called lower right-hand Dini derivate. For $V \in C^0$, if W is a continuous real-valued function and $\dot{V}(t,x) \leq W(t,x)$ for one of the Dini derivates, then the inequality will hold for all four Dini derivates so it makes no difference which Dini derivate is used. In our development "$\liminf, \tau \to +0$" is more convenient than any other derivate.

If V is C^0, then $V'(t,x)$ does not depend on the particular solution chosen and in fact $V'(t,x) = \dot{V}(t,x)$. If V is not a C^0 function, $V'(t,x)$ does depend on the particular solution through (t,x) so when we write $V'(t,\phi(t))$, it will be implicit that V' is evaluated with respect to ϕ. We have found examples with the following behaviors when V is only continuous:

(1) $\dot{V} \equiv 0$ on U and yet there exists a solution ϕ such that $V'(t,\phi(t)) \equiv 1$ for $t \in R$ (or we may have $V'(t,\phi(t)) \equiv -1$ for $t \in R$).

(2) $\dot{V} \equiv 0$ on U and yet there exists a point x_0 such that for every solution ϕ of (E) with $\phi(0) = x_0$, we have $V'(t,\phi(t)) = 1$ for all $t \leq 0$ (or we may have $= -1$ for all $t \leq 0$).

(3) solutions of (E) unique and $V'(t,\phi(t)) \leq 0$ for all t and all solutions ϕ and yet there exists a solution ϕ such that $\dot{V}(t,\phi(t)) = 1$ for all $t \in R$.

Because of the above examples Theorem 1, which is apparently new, is perhaps the best result that can be obtained when V is not C^o. V is lower-semi-continuous (l.s.c.) if $\liminf_{(\tau,\xi)\to(t,x)} V(\tau,\xi) \geq V(t,x)$.

Theorem 1. *If* $W : U \to R$ *is continuous,* V *is* l.s.c., *and* $\dot{V}(t,x) \leq W(t,x)$ *for all* $(t,x) \in U$, *then for each* $(t_o,x_o) \in U$ *there is a solution* $\phi = \phi(\cdot;t_o,x_o)$, *such that for all* $t \geq t_o$ *in the domain of* ϕ,

$$V(t,\phi(t)) - V(t_o,\phi(t_o)) \leq \int_{t_o}^{t} W(s,\phi(s))ds , \tag{1.1}$$

and if V *is continuous, then* $V'(t,\phi(t)) \leq W(t,\phi(t))$.

One main purpose of this note, however, is to introduce another derivate $\overset{*}{V}$ of V which allows nearly the entire Liapunov theory of C^o Liapunov functions to hold for V which need only be l.s.c. Let $|y|$ denote the norm of $y \in R^n$. Define

$$\overset{*}{V}(t,x) = \liminf_{\substack{\tau\to+0 \\ |y|\to 0}} \tau^{-1}[V(t+\tau,\ x+\tau y+\tau f(t,x)) - V(t,x)]$$

From the definition it follows that $\overset{*}{V} \leq \dot{V}$ and $\overset{*}{V} \leq V'$ (calculated along any solution). We can prove Theorem 1 is true with $\dot{V}$ replaced by $\overset{*}{V}$. From this and $\overset{*}{V} \leq V'$ we have

THEOREM 2. *Let* $W : U \to R$ *be continuous and let* V *be* l.s.c. *Then the following two conditions are equivalent:*

(2.1) *For all* $(t_o,x_o) \in U$, *there exists a solution* $\phi = \phi(\cdot;t_o,x_o)$

such that (1.1) *holds for all* $t \geq t_o$ *in the domain of* ϕ .

$$(2.2) \qquad \overset{*}{V}(t,x) \leq W(t,x) \quad \text{for all} \quad (t,x) \in U .$$

Corollary. If solutions are unique and $V : U \to R$ *is continuous,* $\overset{*}{V}(t,x) \leq 0$ *for all* $(t,x) \in U$ *iff for every solution* ϕ , $V(t,\phi(t))$ *is a monotonically decreasing function for* t *in the domain of* ϕ .

Theorem 1 follows immediately from Theorem 2 , and it is not difficult to derive the corollary from Theorem 2.

Sketch of proof of Theorem 2: For any $(t_o,x_o) \in U$ and $\varepsilon > 0$ we may choose τ, y such that, $0 < \tau < \varepsilon$, $|y| < \varepsilon$, and the line segment between (t_o,x_o) and $(t_1,x_1) = (t_o + \tau,\ x_o + \tau y + \tau f(t_o,x_o))$ lies in U and

$$\tau^{-1}[V(t_o + \tau,\ x_o + \tau y + \tau f(t_o,x_o)) - V(t_o,x_o)] - W(t_o,x_o) < \varepsilon .$$

By a sequence of such choices, (t_o,x_o) , (t_1,x_1), ..., (t_i,x_i), ..., we may construct a piece-wise linear "approximate" solution. The sequence $\{(t_i,x_i)\}$ can be chosen so as not to have a cluster point in U . When the approximate solutions are properly chosen some subsequence converges to a solution of (E) satisfying $V'(t ,\phi(t)) \leq W(t,\phi(t))$ for $t \geq t_o$.

Application. The use of $\overset{*}{V}$ rather than $\dot{V}$ will often allow better theorems with simpler proofs, particularly for converse theorems, as in the following application.

We assume now that for some $\eta > 0$ the set $D_\eta = \{(t,x) : |x| \leq \eta$ and $t \geq 0\} \subset U$. Following Strauss [3] , we assume solutions of (E) are unique and we define

Definition. 0 is $\mathcal{L}^p$*-stable* (for (E)) if 0 is stable and if for all $t_o \geq 0$ there exists a $\delta = \delta(t_o) > 0$ such that

$$\int_{t_o}^{\infty} |\phi(t;t_o,x_o)|^p \, dt < \infty \quad \text{for all} \quad |x_o| < \delta \ .$$

Theorem 3. *For any* $p > 0$ *the following are equivalent.*

(3.1) *There exists* $\rho > 0$ *and a lower-semi-continuous positive definite function* $V : D_\rho \to [0,\infty)$ *such that* $V(t,0) \equiv 0$ *and for some* $c > 0$

$$\overset{*}{V}(t,x) \leq c|x|^p \quad \textit{for} \quad (t,x) \in D_\rho \ .$$

(3.2) 0 *is* $\mathcal{L}^p$ *stable* .

Theorem 3 is essentially due to Strauss [3] *except* that he did not have available Theorems 1 and 2. He showed that *if* $V \in C^0$, then (3.1) implies (3.2) , and to conclude (3.1) and $V \in C^0$ he had to assume (3.2) and $f \in C^1$ and

(3.3) for some ρ there exists a function $\psi \in \mathcal{L}^p$ on $[0,\infty)$ such that

$$|\theta(t + t_o,t_o,x_o)| \leq \psi(t) \ , \quad \text{for} \quad (t_o,x_o) \in D_\rho \quad t \geq 0 \ ,$$

where θ is the matrix of partial derivatives $\phi_{x_o}(t;t_o,x_o)$.

BIBLIOGRAPHY

[1] H. OKAMURA, Condition nécessaire et suffisante remplie par les équations différentielles ordinaires sans points de Peano, Mem. Coll. Sci., Kyoto Imperial Univ., Series A, 24(1942), 21-28.

[2] TARO YOSHIZAWA, Stability theory by Liapunov's second method, Math. Soc. of Japan, 1966.

[3] AARON STRAUSS, Liapunov functions and L^p solutions of differential equations, Trans. Amer. Math. Soc., 119(1965), 37-50.

CHARACTERIZING SOLUTIONS OF THE PONTRIAGIN MAXIMUM PRINCIPLE

by

Arrigo Cellina

§1

Let Γ be a multi-valued mapping from a topological space X to a topological space Y and let x^o be a point of X. We shall say that Γ is upper semi-continuous (u.s.c.) at x^o if, for every open set G containing the set Γx^o, there exists a neighborhood $U(x^o)$ such that

$$x \in U(x^o) \Longrightarrow \Gamma x \subset G .$$

For $k > 0$, the closed k-neighborhood of a set $A \subset E^n$ is indicated by $N(A,k)$. $S(E^n)$ is the set of subsets of E^n.

Let us introduce the real variables $x = (x_1 \ldots x_n) \in E^n$, $u = (u_1 \ldots u_r) \in E^r$, t = time, the sets $X \subset E^n$, $Y \in \text{comp}(E^r)$ and the mapping with values in E^n : $f(x,u) = (f_1(x,u), \ldots, f_n(x,u))$ defined on $X \times Y$. It is assumed that $f(x,u)$ is continuous with respect to both the arguments and continuously differentiable with respect to x.

Let $U \in \text{comp}(E^r)$, $U \subset Y$; we shall call a *control system* a couple composed of a differential equation

1.1 $$\dot{x} = f(x,u)$$

and a set of controls

1.2 $$u \in U .$$

A function $u(\cdot)$ is called *admissible* on $[t_o,t_f]$ if

$u(\cdot)$ is measurable on $[t_o,t_f]$

$u(t) \in U$ for almost all t in $[t_o,t_f]$.

An optimum problem for a control system is: given an initial condition $x(0) = \xi^o$ and a target set T , find an admissible control $u(\cdot)$ such that, if $x(\cdot)$ is the trajectory corresponding to $u(\cdot)$ on $[t_o,t_f]$, then $x_n(t_f)$ is minimum with respect to all other admissible controls.

It is required that the function $f(x,u)$ does not depend on x_n , and that the target set is a smooth manifold of dimension less than n.

Let us introduce the n-dimensional vector p and consider the mapping $H : E^n \times E^n \times E^r \to E^1$

$$H(x,p,u) = p'f(x,u) .$$

We can now state the

1.3 *Pontriagin Maximum Principle* [1] . Let $u(t)$, $t_o \leq t \leq t_f$, be an admissible control; for the corresponding trajectory $x(\cdot)$, $x(0) = \xi^o$ to be optimal, it is necessary that the following conditions be fulfilled:

1.4 There exists a vector valued function, continuous and non vanishing, $p(\cdot) = (p_1(\cdot), \ldots, p_n(\cdot))$, a solution of the system of differential equations

$$\dot{p} = -f_x(x,u)'p$$

such that

$$H(x(t),\ p(t),\ u(t)) = \max_{u \in U} H(x(t),\ p(t), u)$$

for every t, $t_o \leqslant t \leqslant t_f$;

1.5 if t_f is fixed, at t_f a set of relations

$$p_n(t_f) \leqslant 0$$

$$v_i(p,x) = 0 \qquad\qquad i = 1, \ldots, (n-1)$$

have to be satisfied, where $v_i(\cdot,\cdot)$ are continuous functions, homogeneous in p , depending on the specific problem.

1.6 if t_f is free, the preceding relations have to be satisfied in addition to

$$\max_{u \in U} H(x(t_f),\ p(t_f), u) = 0 \quad .$$

In what follows, we shall call "extremals" the function couples

$(x(\cdot), p(\cdot))$ satisfying condition 1.4 of the Maximum Principle, and "extremals satisfying the boundary conditions" (extremals s.b.c.) those satisfying also conditions 1.5 and 1.6.

For every couple (x,p) in $X \times E^n$ define the mapping $H^o : E^n \times E^n \to E^1$

1.7 $$H^o(x,p) = \max_{u \in U} H(x,p,u) \quad .$$

Since $H(x,p,u)$ is continuous and U is compact, for every (x,p), H^o exists.

For every (x,p) in $X \times E^n$ there then exists one or more vectors u such that $H(x,p,u) = H^o(x,p)$. We are allowed then to define a multi-valued mapping $M : E^n \times E^n \to S(E^r)$ in the following way

1.8 $$u \in M(x,p) \quad \text{if} \quad H(x,p,u) = H^o(x,p) \quad .$$

For a given (x,p), $M(x,p)$ is the set of all $u \in U$ that maximize the Hamiltonian function.

1.9 *Proposition.* $H^o(x,p)$ *is continuous.*

1.10 *Lemma.* *The multi-valued mapping* $M(x,p)$ *is upper semi-continuous.*

Let us define the mapping $c : E^m \times E^m \times E^r \to E^n \times E^n$

1.11 $$c(x,p,u) = \begin{pmatrix} f(x,u) \\ \\ -f_x(x,u)'p \end{pmatrix}$$

and consider for each (x,p) the set $R(x,p)$ obtained when u describes the set $M(x,p)$:

1.12 $$R(x,p) = \{y = c(x,p,u) \; ; \; u \in M(x,p)\} \; .$$

Then we have

1.13 *Lemma. The set valued mapping* $R(x,p)$ *is upper semi-continuous.*

By means of the set valued mapping $R(x,p)$ we have defined on $X \times E^n$ an orientor field.

We want to consider the multi-valued differential equation associated with such a field:

1.14 $$\begin{pmatrix} \dot{x} \\ \dot{p} \end{pmatrix} \in R(x,p) \; .$$

Then the main result follows:

1.15. *Theorem. The set of solutions of the multi-valued differential equation is equivalent to the set of functions* $(x(\cdot),p(\cdot))$ *satisfying condition* 1.4 *of the Maximum Principle.*

For the proof of the preceding theorem and lemmas no convexity hypothesis is needed.

We have given a unified description of the set of extremals by means of the single multi-valued differential Equation 1.14, while elsewhere this set of extremals is described by "splitting" the phase space into different regions and considering various differential equations, or classes of differential equations, in the various regions. This spitting procedure makes it hard to keep all the continuity properties of the trajectories of the field, which are instead preserved in our approach.

Moreover, in the classical formulation of the Pontriagin Maximum Principle, an extremal is defined only in the interval $[t_o, t_f]$, where t_f is the time at which the phase point hits the target; while in our approach extremals can be considered in their maximal domain of existence. This property makes it much easier to handle sequences of extremals.

In this section we are concerned with the continuous dependence (or, better, semi-continuous dependence, as it will be shown) of extremals satisfying the boundary conditions, both on the initial data and on parameters.

Let us begin with the following problem: let there be given an optimal control problem and suppose that we know at least one extremal satisfying the boundary conditions, for every initial point in an open region $0 \subset E^n$: can we extend our knowledge to the boundary of 0 ? In other words, given an arbitrary point on the boundary of 0 , can we be sure that there will be at least one extremal through it, satisfying the boundary conditions, and will it be possible to obtain at least one such extremal by a limit process?

That the answer to the first question is positive is probably known under certain conditions; this is anyway implied in the answer to the second question that we are going to give, using the semi-continuity properties of the set of solutions of a multi-valued differential equation.

In order to obtain the results of this section some hypotheses are needed. Precisely, in the sequel we shall admit the following

2.1 *Fundamental Hypothesis.* The set $R(x,p)$ is convex for $(x,p) \in X \times E^n$.

2.2 *Remark.* For $(x,p) \in X \times E^n$ let us call $C(x,p)$ the subset of $E^n \times E^n$ described by $c(x,p,u)$ when u describes the whole set U, instead of being restricted to $M(x,p)$. Since for checking Hypothesis 2.1 it is necessary to maximize the Hamiltonian, it may be convenient to assume instead the following more immediate (but more restrictive)

2.3 *Optional Hypothesis.* For $(x,p) \in X \times E^n$, the set $C(x,p)$ is convex.

The proof that

$$C(x,p) \text{ convex} \Longrightarrow R(x,p) \text{ convex}$$

is straightforward.

2.4 *Remark.* It is classical to assume, instead of the convexity of $R(x,p)$, the convexity of $F(x)$, where $F(x)$ is the subset of E^n described by $f(x,u)$ when u ranges over U.

It is possible to realize that, if the function $f(x,u)$ has the form

$$f(x,u) = g(x) + A(x)h(u) ,$$

where $A(x)$ is a diagonal matrix whose elements are functions of x , then the convexity (in E^n) of $F(x)$ implies the convexity (in $E^n \times E^n$) of $C(x,p)$.

On the other hand we want to point out that it is possible to give examples satisfying the condition "$R(x,p)$ convex" but *not* the condition "$F(x)$ convex."

Hypothesis 2.3 in addition to compactness and upper semi-continuity supplies the set-valued mapping $R(x,p)$ with all the regularity properties required by the classical theory of multi-valued differential equations. We are then able to state the following

2.5 *Proposition.* For every $\zeta^o \in X \times E^n$ there exists at least one solution of Equation 1.14 satisfying the initial condition $\zeta(0) = \zeta^o$. Such a solution can be continued for all t or until reaching the boundary of $X \times E^n$.

More, if by $S(\zeta^o)$ we denote the set of solutions $\zeta(\cdot)$, defined on $[t_o,t_f]$, such that $\zeta(0) = \zeta^o$, the multi-valued mapping $S : X \times E^n \to C_{E^{2n}}[t_o,t_f]$ is u.s.c.

This next theorem follows:

2.6 *Theorem. Let* $I \in \mathrm{comp}(E^1)$*; let* $\Omega \subset E^n$ *such that for a* $k > 0$, $N(\Omega,k) \subset X$; $N(\Omega,k) \in \mathrm{comp}(E^n)$*; let* $\{\xi^m\} \subset \Omega$, $\xi^m \to \xi^o$.

Suppose that for $m = 1, 2, \ldots$ *there exist absolutely continuous functions* $x^m(\cdot)$ *, defined on* $I^m = [t_o, t_f^m]$, $I^m \subset I$, $x^m(t) \in \Omega$, $t \in I^m$ *, such that* $x^m(\cdot)$ *are extremals* s.b.c. *, and* $x^m(0) = \xi^m$.

Then it is possible to construct, by a limit process, at least one extremal s.b.c. $x^o(\cdot)$ *with initial data* $x^o(0) = \xi^o$.

We are now going to extend the results on the continuous dependence on the initial data as in the classical theory of ordinary differential equations to the continuous dependence on parameters.

Suppose a class of control problems is given, described by the set of differential equations

2.7 $$x = f(x,u,\varepsilon) ,$$

where ε is an s-dimensional vector belonging to a region $L \in \mathrm{comp}(E^s)$. We assume that f is continuously differentiable also with respect to the parameter ε for $\varepsilon \in L$.

It is possible to transform the problem of the continuous dependence on parameters to the problem of the continuous dependence on the initial data, simply adding to the system 2.7 the system of differential equations

2.8 $$\frac{d\varepsilon}{dt} = 0 .$$

Let us introduce the $(n+s)$-dimensional vectors $g = (f,0)$, $y = (x,\varepsilon)$, $q = (p,\eta)$ and consider the system

$$\dot{y} = g(y,u)$$

$$\dot{q} = -g_y(y,u)'q \quad .$$

We shall assume the following

2.9 *Fundamental Hypothesis.* For $y \in X \times L$, $q \in E^{n+s}$, the set $R(y,q)$ is convex.

Then the following theorem can be proved:

2.10 *Theorem. Let* $I \in \text{comp}(E^1)$*; let* $\Omega \subset E^n$ *such that for a* $k > 0$, $N(\Omega,k) \subset X$, $N(\Omega,k) \in \text{comp}(E^n)$*; let* $\{\xi^m\} \in \Omega$, $\xi^m \to \xi^o$, $\{\varepsilon^m\}$ L , $\varepsilon^m \to \varepsilon^o$. *Suppose that for* $m = 1, 2, \ldots$ *there exist absolutely continuous functions* $x^m(\cdot)$, *defined on* $I^m = [t_o, t_f^m]$, $I^m \subset I$, *such that* $x^m(\cdot)$ *are extremals satisfying the boundary conditions of the problem with parameters* ε^m *and initial data* $x^m(0) = \xi^m$.

Then it is possible to construct, by a limit process, at least one extremal $x^o(\cdot)$, *satisfying the boundary conditions, of the problem with parameters* ε^o *and initial data* $x^o(0) = \xi^o$.

REFERENCES

[1] L. PONTRIAGIN *et al*, The Mathematical Theory of Optimal Processes, Interscience, New York, 1962.

LIAPUNOV FUNCTIONS AND THE EXISTENCE OF SOLUTIONS TENDING TO 0

by

James A. Yorke

We construct a Liapunov theory for the non-linear equation

$$(E) \qquad \dot{x} = F(x)$$

(where $x \in R^n$ and $F : R^n \to R^n$ is a C^1 function) with which we can guarantee the *existence* of at least one non-trivial solution ϕ with some special property such as (i) $\phi(t) \to K$ as $t \to \infty$ for K some compact set, (ii) ϕ remains bounded on $[0,\infty)$, or (iii) $|\phi(t)| \to \infty$ as $t \to \infty$. Theorem 2 classifies all solutions of (E) using hypotheses similar to those of usual Liapunov theory.

Our main tool is the "Liapunov" function V. From now on we assume $V: R^n \to [0,\infty)$ is a C^2 function. We will also assume $F \in C^1$ so that we may define for (E)

$$\dot{V}(x) = \langle \text{grad } V(x) , F(x) \rangle \quad \text{and}$$

$$\ddot{V}(x) = \langle \text{grad } \dot{V}(x) , F(x) \rangle$$

where $\langle \cdot , \cdot \rangle$ denotes the inner product in R^n. We will let $\phi(t,x)$ denote the solution of (E) such that $\phi(0,x) = x$.

Therefore (holding x fixed) we have

$$\dot{V}(\phi(t,x)) = \frac{d}{dt} V(\phi(t,x)) \quad \text{and}$$

$$\ddot{V}(\phi(t,x)) = \frac{d^2}{dt^2} V(\phi(t,x)).$$

The function $\ddot{V}$ has been investigated by two authors, Kudaev [2,3] and Chu [1], and in a less general setting by Nohel [4].

We say K is an *invariant* set if $x \in K$ implies $\phi(t,x) \in K$ for all t. We assume for simplicity that all solutions of all equations considered are defined on all of R^1. For any set S we write bnd S for *the boundary of S*, and $|\cdot|$ will denote any convenient norm on R^n.

Theorem 1. *Let* U_r *be a bounded connected component of* $\{x \mid V(x) < r\}$. *Assume*

(i) *for all* $x \in \text{bnd } U_r$

$$\dot{V}(x) = 0 \quad \textit{implies} \quad \ddot{V}(x) > 0;$$

(ii) *there exists* $y \in \text{bnd } U_r$ *such that* $\dot{V}(y) \leq 0$;

(iii) U_r *has a nonempty compact invariant subset* K .

Then

(iv) *there exists* $z \in \text{bnd}\, U_r$ *such that*

$$\phi(t,z) \in U_r \qquad \textit{for all} \quad t > 0.$$

If instead of (ii) *and* (iii) *we assume that* (i) *holds and*

(v) *the set* $\{y \in \text{bnd}\, U_r \mid \dot{V}(y) > 0\}$ *is non-empty and not connected,*

then

(vi) *the set* $\{ z \mid \phi(t,z) \in U_r$ *for all* $t > 0\}$ *has dimension at least* $n - 1$.

The above theorem guarantees the existence of a bounded solution. Of particular interest in the next theorem is the case $K = \{0\}$ and the existence of z_1 when the solutions through some points do *not* tend to 0 as $t \to \infty$.

Our main condition is that (for $x \in \text{bnd}\, U_r$ or for $x \in R^n - K$) *if* $\dot{V}(x) = 0$, then $\ddot{V}(x) > 0$. This condition may be compared with the conditions for Liapunov-type results which usually require that some inequality or equality holds everywhere in a specified set. None the less, the conclusion in Theorem 2 is not much weaker in that Theorem 2 still gives a complete classification of the solutions. Note that if condition (iv) of Theorem 2 is not satisfied we are reduced to the usual theories with $\dot{V} < 0$ or $\dot{V} > 0$.

Theorem 2. *Let* K *be a compact invariant set and let* V *be chosen so that*

(i) $V(x) = 0$ *for all* $x \in K$, *and*

$V(x) \geq 0$ *for all* $x \in R^n$;

(ii) $V(x) \to \infty$ *as* $|x| \to \infty$;

(iii) $\dot{V}(v) = 0$ *implies* $\ddot{V}(x) > 0$ *for all* $x \in R^n - K$.

(iv) $\dot{V}(y) = 0$ *for some* $y \in R^n - K$.

Then there exist points z_1 *and* z_2 *in* $R^n - K$ *such that*

(v) $\phi(t,z_1) \to K$ *as* $t \to \infty$

$|\phi(t,z_1)| \to \infty$ *as* $t \to -\infty$,

(vi) *and* $|\phi(t,z_2)| \to \infty$ *as* $t \to \infty$

$\phi(t,z_2) \to K$ *as* $t \to -\infty$.

(vii) *For all* $z \in R^n - K$, *either* $\phi(t,z)$ *behaves as in* (v) *or* (vi) *or* $|\phi(t,z)| \to \infty$ *as* $|t| \to \infty$, *and, furthermore, if* $\phi(t,z_o) \to \infty$ *as* $t \to +\infty$ *(or* $t \to -\infty$*)*, *for some* z_o, *then* $\phi(t,z) \to \infty$ *as* $z \to z_o$ *and* $t \to \infty$ *(or* $t \to -\infty$*)*.

Kudaev [2] has certain results closely related to Theorem 2 for $K = \{0\}$. He has also developed [3] a classification scheme for the cases in which $\ddot{V}$ assumes both positive and negative values on the set

$$\dot{V}^o = \{x \mid \dot{V}(x) = 0\} \quad .$$

All his theorems, however, make several extra assumptions on the geometric nature of $\dot{V}^o$. We examine only the case in which $\ddot{V}$ is strictly positive on $\dot{V}^o - K$ because examples suggest that this is the most important case.

Example. Let K be the origin in Theorem 2, and suppose there exists an $n \times n$ matrix A with n distinct eigenvalues, none having zero real part, such that $Ax \equiv F(x)$. Then there exists an inner product $\langle \cdot , \cdot \rangle$ on R^n such that for $V(x) = \langle x,x \rangle$ conditions (i), (ii), and (iii) of Theorem 2 hold. Then (iv) holds iff there is at least one eigenvalue with negative real part and at least one with positive real part.

Theorem 2 is sufficiently precisely stated that a complete converse is possible. We see in Theorem 3 that if all the "geometric" properties of the solutions hold then there must exist a function V as is assumed in Theorem 2.

Theorem 3 *(Converse for theorem 2). Let* K *be a compact, invariant set. Assume* (vii) *from Theorem 2 is satisfied for* (E) *and that there exist* z_1 *and* z_2 *such that* (v) *and* (vi) *are satisfied.*

Then there exists a C^2 *function* $V : R \to R$ *satisfying* (i) - (iv) .

BIBLIOGRAPHY

[1] H. S. CHU, On the n-th derivatives of a Liapunov function, to appear.

[2] M. B. KUDAEV, The use of Ljapunov functions for investigating the behavior of trajectories of systems of differential equations, Soviet Math. 3(1962), 1802-4, (trans. from Doklady Akad. Nauk. SSSR).

[3] M. B. KUDAEV , Classification of higher-dimensional systems of ordinary differential equations by the method of Lyapunov functions, Differential Equations, 1(1965), 263-9, (trans. from Differentsial'nye Uravneniya).

[4] JOHN A. NOHEL, Problems in qualitative behavior of solutions of nonlinear Volterra equations, Nonlinear Integral Equations, edited by P. M. Anselone, University of Wisconsin Press (1964), 191-214.

[5] O. PERRON, Ueber Stabilität und asymptotisches Verhalten der Integrale von Differentialgleichungssystemen, Math. Zeit. 29(1929), 129-160.

FUNDAMENTAL MATRIX IN LINEAR FUNCTIONAL DIFFERENTIAL EQUATIONS

by

Junji Kato

Recently, many authors have dealt with linear functional differential equations and obtained many results similar to those in linear ordinary differential equations. It is also true in functional differential equations that the set of solutions of a linear system is a linear space. However, unfortunately, this linear space is, in general, of infinite dimension.

On the other hand, Halanay [1] has shown that for a non-homogeneous linear system

$$\dot{x}(t) = F(t,x_t) + f(t) .$$

the solution through ϕ at $t = t_o$ (i.e., $x(t_o - s) = \phi(-s)$) is given by

$$x(t;\phi,t_o) = y(t;\phi,t_o) + \int_{t_o}^{t} Y(t;s)f(s)ds ,$$

where $y(t;\phi,t_o)$ is a solution of the homogeneous linear system

(1) $$\dot{y}(t) = F(t,y_t)$$

through ϕ at $t = t_o$ and $Y(t;s)$ is a square-matrix solution of (1) such that

$$Y(s;s) = E\ , \quad Y(s+\theta;s) = 0 \quad \text{for} \quad \theta < 0\ .$$

To prove this, Halanay utilized an adjoint system of (1), which he defined so that ${}^{T}Y(s;t)$ is the matrix solution of the adjoint system. Comparing this with the results in ordinary differential equations, $Y(t;s)$ seems to play a role similar to that of the fundamental matrix. In this paper, we shall call $Y(t;s)$ the *fundamental matrix* of the system (1) and consider its properties.

Let C be the space of all continuous R^n-valued functions defined on $[-r,0]$ for a given constant $r > 0$, and let $||\phi||$ be the norm in C given by

$$||\phi|| = \sup\{|\phi(\theta)|;\ 0 \geqq \theta \geqq -r\}\ .$$

For a continuous R^n-valued function $x(s)$, x_t denotes the function in C such that

$$x_t(\theta) = x(t+\theta) \quad \text{for} \quad \theta \in [-r,0]$$

and $\dot{x}(t)$ denotes the right-hand derivative. Consider the system

$$\text{(2)} \qquad \dot{x}(t) = F(t,x_t)\ ,$$

and assume that $F(t,\phi)$ is continuous in (t,ϕ) and linear in ϕ on $[0,\infty) \times C$.

First of all, we can prove the following lemma.

Lemma 1 . *Under the assumption on* $F(t,\phi)$, *there exists a continuous function* $L(t)$ *such that*

$$|F(t,\phi)| \leq L(t) \, \|\phi\|$$

for all $(t,\phi) \in [0,\infty) \times C$.

Therefore, there arise no difficulties in the existence and uniqueness problem of the fundamental matrix of the system (2), though the initial function of the fundamental matrix is not continuous (for example, see [3]).

Let $X(t;s)$ be the fundamental matrix of system (2). Then, we have the following lemma.

Lemma 2 . *For any* $\phi \in C$, *any* $\varepsilon > 0$ *and any* $t \geq r$, *there exists a positive integer* m , *a sequence of real numbers* $\{\theta_i\}$,

$$-r \leq \theta_0 < \theta_1 \quad \cdots < \theta_m \leq 0 ,$$

and a sequence of constant vectors $\{c_i\}$, $c_i \in R^n$ *for* $i = 0, 1, \ldots, m$, *such that*

$$\|\phi - \sum_{i=0}^{m} X_t(t + \theta_i)c_i\| < \varepsilon , \tag{3}$$

where $X_t(s)$ *is the segment of* $X(\tau;s)$, *that is,*

$$X_t(s)(\theta) = X(t+\theta;s) .$$

Furthermore, if ϕ *satisfies*

$$|\phi(\theta) - \phi(\theta')| \leq L|\theta - \theta'|$$

for all θ, $\theta' \in [-r,0]$, *then the above sequences can be chosen so that*

$$\overline{\lim_{\varepsilon \to 0}} \sum_{i=0}^{m} |c_i| \leq (2r \max\{M(t),L\} + 1) \|\phi\| ,$$

where

$$M(t) = \sup\{ L(u)\exp[\int_s^u L(\tau)d\tau] ; t \geq u \geq s \geq t - r\}$$

and $L(t)$ *is the function given by Lemma* 1.

By the first part of Lemma 2, if $H(t,\phi)$ is continuous in (t,ϕ) and is linear in ϕ and if

$$H(t,X_t(s)c) = 0$$

for the fundamental matrix $X(t;s)$ of a linear system and for all constant vectors $c \in R^n$, then

$$H(t,\phi) = 0$$

for any $\phi \in C$. Thus we have the following theorem.

Theorem 1 . *Consider the systems* (2) *and*

(4) $$\dot{x}(t) = G(t,x_t) ,$$

where $G(t,\phi)$ *is continuous in* (t,ϕ) *and linear in* ϕ . *Let* $X(t;s)$, $Y(t;s)$ *be the fundamental matrices of the systems* (2) *and* (4) *respectively, and suppose that for an interval* $[a,b]$

$$X(t;s) = Y(t;s) \quad \textit{for all} \quad t,s \in [a,b],\ t \geq s .$$

Then we have

$$F(t,\phi) = G(t,\phi)$$

for all $(t,\phi) \in [a + r,b] \times C$.

For an autonomous linear system

$$\dot{x}(t) = \int_{-r}^{0} [d\ \eta(\theta)]x(t + \theta) ,$$

Hale [2] has defined the adjoint system by

(5) $$\dot{y}(t) = - \int_{-r}^{0} [d^T\eta(\theta)]\ y(t - \theta) ,$$

where $\eta(\theta)$ is a matrix function of bounded variation and the integrations are in the sense of Stieltjes. On the other hand, Halanay [1] has defined the adjoint system of the system

$$\dot{x}(t) = \int_{-r}^{0} [d\eta(t,\theta)]\, x(t+\theta)$$

by

$$\frac{d}{dt}\left[y(t) + \int_{-\infty}^{0} {}^{T}\eta(t-\theta,\theta)\, y(t-\theta)d\theta\right] = 0 . \tag{6}$$

Here, he assumes that $\eta(t,\theta)$ is continuous in t uniformly in θ and is of bounded variation in θ, and he sets $\eta(t,\theta) = \eta(t, -r)$ for all $\theta < -r$ and $\eta(t,0) = 0$.

However, each of them showed that the fundamental matrix $Y(t;s)$ of his adjoint system is given by

$$Y(t;s) = {}^{T}X(s;t) , \quad s \geqq t ,$$

for the fundamental matrix $X(t;s)$ of the original system. Therefore, in the case where $\eta(t,\theta) \equiv \eta(\theta)$, Theorem 1 shows that the system (5) and (6) must be equivalent to each other. In other words, if $\eta(\theta)$ is of bounded variation and satisfies

$$\eta(\theta) = 0 \quad \text{for} \quad \theta \geqq 0 , \quad \eta(\theta) = \eta(-r) \quad \text{for} \quad \theta \leqq -r$$

and if $\phi(\theta)$ is a continuous function which is defined on $(-\infty,a)$ for a > 0 and satisfies

$$\int_{-\infty}^{0} |\phi(\theta)|d\theta < \infty ,$$

then we have

$$\int_{-r}^{0} [d\eta(\theta)]\phi(\theta) = \frac{d}{dt} [\int_{-\infty}^{0} \eta(\theta)\phi(\theta - t)d\theta] \Big|_{t=0}$$

$$= \frac{d}{dt} [\int_{-\infty}^{-t} \eta(t + \theta)\phi(\theta)d\theta] \Big|_{t=0}$$

$$= \frac{d}{dt} [\int_{-\infty}^{0} \eta(t + \theta)\phi(\theta)d\theta] \Big|_{t=0}$$

that is, roughly speaking, we can commute the order of the integration and the differentiation in the Stieltjes integration.

Theorem 2 . *For any* $(t_o,\phi) \in [r,\infty) \times C$, *a solution* $x(t;\phi,t_o)$ *of the system* (2) *satisfies*

$$|x(t;\phi,t_o)| \leq (1 + 2rM^*(t_o))\Phi(t;t_o)B(t_o) \, ||\phi||$$

for all $t \geq t_o$, where

$$M^*(t_o) = \sup\{M(s)\ ;\ t_o \leq s \leq t_o + r\} \quad ,$$

$$B(t_o) = \exp\left[\int_{t_o}^{t_o+r} L(s)ds\right] \ ,$$

$$\phi(t;t_o) = \begin{cases} \sup\{|X(t;s)|\ ;\ t_o + r \geq s \geq t_o\} & \textit{for} \quad t \geq t_o + r \\ 1 & \textit{for} \quad t_o + r > t \geq t_o \ . \end{cases}$$

This theorem can be proved by the second part of Lemma 2 and by the fact that $x(t;\phi,t_o)$ satisfies

$$|x(t;\phi,t_o)| \leq B(t_o)\ ||\phi|| \quad ,$$

$$|x(t;\phi,t_o) - x(t';\phi,t_o)| \leq M(t_o)\,|t - t'|$$

on the interval $[t_o,t_o + r]$.

From this theorem, we can obtain the following.

Corollary.

(i) *If* $\Phi(t;t_o)$ *is bounded on* $[t_o,\infty)$ *for any* $t_o \geq 0$ *, then all solutions of the system* (2) *are bounded.*

(ii) *If* $\Phi(t;t_o)$ *tends to zero as* $t \to \infty$ *for all* $t_o \geq 0$ *, then each solution tends to zero as* $t \to \infty$.

(iii) *If* $L(t)$ *can be chosen to be constant and if* $X(t;t_o)$ *is bounded uniformly in* t_o *, then solutions of the system* (2) *are uniformly bounded.*

(iv) *If* $L(t)$ *can be chosen to be constant and if* $X(t;t_o)$ *tends to zero as* $t \to \infty$ *uniformly in* t_o *, then all solutions tend to zero as* $t \to \infty$ *uniformly in* t_o .

(v) *If* $F(t,\phi)$ *in the system* (2) *is independent of* t *and if each solution of the system* (2) *is bounded, then solutions of* (2) *are uniformly bounded.*

It is expected that if all $X(t;t_o)$ are bounded or tend to zero as $t \to \infty$, then $\Phi(t;t_o)$ is bounded or tends to zero as $t \to \infty$. If we can prove this, then the boundedness of solutions of a linear system implies the equiboundedness, and hence, for a linear system boundedness and stability are equivalent. However, this question is still open.

REFERENCES

[1] A. HALANAY, Differential Equations, Academic Press, 1966.

[2] J. K. HALE, Linear functional-differential equations with constant coefficients, Contr. Diff. Eqs., 2(1964), 291-317.

[3] N. N. KRASOVSKII, Some Problems in the Theory of Stability of Motion, Stanford Univ. Press, 1963.

ASYMPTOTIC STABILITY FOR FUNCTIONAL DIFFERENTIAL EQUATIONS

by

James A. Yorke

For non-linear one-dimensional functional differential equations

(F) $$\dot{x} = F(t, x_t)$$

one can frequently (almost by inspection) determine whether the 0 solution is asymptotically stable and give a region of attraction. Rather than make an inordinate number of assumptions on how $F(t,\phi)$ depends on ϕ ,(F) is rewritten as

(A) $$\dot{x} = -a(t, x_t)\; x(t - r(t, x_t)) \; .$$

Since we are not trying to conclude existence of solutions for (A) we have to make *no continuity assumptions* on $r(t,\phi)$ and $a(t,\phi)$. The examples indicate the generality of this theorem. An advantage of this theorem is that it gives a simple criterion to examine, and no use is made of complicated criteria as in Liapunov theory where Liapunov functions usually cannot be found, even for the most elementary functional differential equations.

This paper was stimulated by a "Research Problem" by Bellman [1] in which he suggested studying in one dimension

$$\dot{x} = -\alpha x(t - r(t)) .$$

Cooke, [2] and [3, p. 167-183], studied this equation where α is an arbitrary constant and $r(\cdot)$ is continuous and $r(t) \to 0$ as $t \to \infty$. He proved that there is a solution ψ such that $\psi(t)e^{\alpha t} \to$ constant $\neq 0$ as t Kato, (see [4,5] and his lectures), has generalized Cooke's studies of this ec to higher dimensions, allowing α to be a constant matrix and with assumptions on r similar to Cooke's. He was able to prove the existence of such solutions even in the particularly difficult case in which α has 0 eigenvalues. V. I. Logunov [8] has also worked on higher order linear systems with the lag function integrable on $[0,\infty)$.

Because of the difficulty of dealing with even simple functional differential equations, the equations are often of necessity unclearly related to the physical or biological situation they should represent. For example population models often have a single lag rather than a functional dependence depending on the entire distribution of the population for a generation. Although the equation's solutions may seem to agree closely with the actual population growth, present-day perturbation theory for functional differential equations cannot account for the agreement. Solutions of functional differential equations frequently behave more regularly than any existing theory predicts. Theorem 1 in this lecture gives one kind of general qualitative behavior that can be predicted for very general equations.

Notation. For a continuous function $\phi : [-q,0] \to R$, let $||\phi|| = \sup_{t \in [-q,0]} |\phi(t)|$. Let $C[t_o,t_1]$ denote the set of continuous functions $\phi : [t_o,t_1] \to R$ and C^q_β be the set of $\phi \in C[-q,0]$ such that $||\phi|| < \beta$. Let $J = (0,\infty)$. We will assume $F : J \times C^q_\beta \to R$ is continuous. If ψ is defined on at least $[t - q,t]$, we will write ψ_t for the function such that $\psi_t(s) = \psi(t + s)$ for $s \in [-q,0]$. Hence $\psi_t \in C[-q,0]$. We say ψ is a solution on $[t_o,t_1)$ of a functional differential equation if ψ is an absolutely continuous real valued function, defined on an interval $[t_o - q, t_1)$, where $\infty \geq t_1 > t_o > t_o - q$, which satisfies the equation almost everywhere on $[t_o,t_1)$. Let $\psi = \psi(\cdot;t_o,\phi)$ denote a solution which satisfies $\psi_{t_o} = \phi$. Note that since our theorems say that all solutions behave in a certain manner, the more general our definition of solution, the better the theorem .

Definition 1 . We say 0 is *uniformly stable* for (F) if for any $\eta > 0$ there exists a $\delta = \delta(\eta)$ in $(0,\eta]$ such that for any $t_o > 0$ and $||\phi|| < \delta$ and any solution $\psi = \psi(\cdot;t_o,\phi)$ we have for all $t > t_o$ in the domain of ψ

$$|\psi(t;t_o,\phi)| < \eta .$$

Definition 2 . We say 0 is *asymptotically stable* for (F) if 0 is uniformly stable and there exists $\eta > 0$ such that for any $t_o > 0$, every solution $\psi = \psi(\cdot;t_o,\phi)$ defined on $[t_o,\infty)$ with $||\phi|| < \eta$ we have

$$\psi(t) \to 0 \qquad \text{as} \quad t \to \infty .$$

It is well known and easy to show that if F is continuous and there exist $B > 0$ and $\eta > 0$ such that $|F(t,\phi)| < B$ when $t > 0$ and $|\phi| < \eta$, and if 0 is uniformly stable for (F), then for $||\phi|| < \delta(\eta)$ and $t_o > 0$, there exists a solution $\psi(\cdot;t_o,\phi)$ defined on all of $[t_o,\infty)$.

Definition 3. For each $\theta \in R$, define $\phi^\theta \in C^q$ such that $\phi^\theta(t) \equiv \theta$. The function $a : J \times C^q \to R$ is *uniformly non-zero near each non-zero constant* if for each θ, $0 < |\theta| < \beta$, there exists $\eta_\theta > 0$ such that if $||\phi - \phi^\theta|| < \eta_\theta$ then $|a(t,\phi)| \geq \eta_\theta$. Note that when a is independent of t, $a(t,\phi) = a(\phi)$, and is continuous, then it is sufficient to have $a(\phi^\theta) \neq 0$ for $\rho > |\theta| > 0$. Furthermore, if $F(t,\phi) = a(t,\phi)\phi(-r(t,\phi))$ then a is uniformly non-zero near each non-zero constant iff F is.

We now state the main theorem. It is important to note that a and r are not assumed to be continuous since, even if F is continuous and $F = a\phi(-r)$, the functions a and r may not be continuous.

Theorem 1. *For some* $\beta > 0$, $q > 0$, $\rho \in [0,q]$, *let* $a : J \times C^q_\beta \to [0,\alpha]$ *and* $r : J \times C^q_\beta \to [0,\rho]$. *Hence,* a *and* r *are bounded by* α *and* ρ.

(1.1) *If* $\alpha\rho \leq 3/2$, *then* 0 *is uniformly stable, and if* $\phi \in C^q$, $||\phi|| < \frac{2}{5}\beta$, $t_o > 0$, *then*

$$|\psi(t;t_o,\phi)| \leq \frac{5}{2}||\phi|| \qquad \textit{for} \quad t \geq t_o .$$

(1.2) *If furthermore* $\alpha\rho < 3/2$ *and* ψ *is defined on* $[t_o,\infty)$, *then*

$$\psi(t) \to \textit{constant} \qquad \textit{as} \quad t \to \infty .$$

(1.3) *If furthermore* $a(\cdot,\cdot)$ *is uniformly non-zero near each non-zero constant* ψ , *then* 0 *is asymptotically stable.*

(1.4) *If* $\alpha\rho < e^{-1}$ *and* $\phi \in C^q$, $||\phi|| < \frac{2}{5}\beta$, $t_o > 0$, *and if no two zeroes of* $\psi = \psi(\cdot;t_o,\phi)$ *on* $[t_o,\infty)$ *differ by less than* ρ , *then* ψ *is non-zero and monotonic on* $[t_o + \rho,\infty)$.

Remarks and Examples. (1) Given F we may define $a(t,\phi)$ as follows

$$a(t,\phi) = \begin{cases} 0 & \text{for } ||\phi||\ F(t,\phi) = 0 \\ F(t,\phi)[\sup\limits_{[-q,0]} \phi]^{-1} & \text{for } ||\phi||\ F(t,\phi) > 0 \\ F(t,\phi)[\inf\limits_{[-q,0]} \phi]^{-1} & \text{for } ||\phi||\ F(t,\phi) < 0 \end{cases}$$

Define $r = r(t,\phi)$ such that

$$\phi(-r) = \begin{cases} \max\limits_{[-q,0]} \phi & \text{for } F(t,\phi) \geq 0 \\ \min\limits_{[-q,0]} \phi & \text{for } F(t,\phi) < 0 \ . \end{cases}$$

Then for any continuous function ψ ,

$$F(t,\psi_t) = a(t,\psi_t)\ \psi_t(-r(t,\psi_t)) = a(t,\psi_t)\ \psi(t - r(t,\psi_t))$$

we have defined $a(\cdot,\cdot)$ such that $a(t,\phi)$ will be positive if ϕ is both positive and negative on $[-q,0]$, or if ϕ has a single sign and

$F(t,\phi)$ has the same sign. In such cases the above definition makes $a(t,\phi)$ as small as possible. However, an alternate choice of a and r, which in practice is almost as good, may be given when $F(t,\phi)$ is defined by a Stieltjes integral. The following intermediate value theorem for the Riemann-Stieltjes integral is useful. See [6, p. 213].

Let ν *be a monotonically increasing function and* g *continuous on* $[t_o,t_1]$. *Then there exists* $r \in [t_o,t_1]$ *such that*

$$\int_{t_o}^{t_1} G(s)d\nu(s) = (\nu(t_1) - \nu(t_o))\, G(r) .$$

(2) Note that if $F(t,\phi) \equiv 0$, then we have $F \equiv ax$ with $a \equiv 0$ and $r \equiv 0$. The conditions of (1.1) and (1.2) are satisfied (letting $r = \rho = 0$) but 0 is not asymptotically stable. Therefore some extra condition (as in (1.3)) is needed to guarantee that 0 is asymptotically stable.

(3) For the equation

$$\dot{x} = -\alpha x(t - \rho)$$

with $\alpha \geq 0$ and $\rho \geq 0$ constant, it is well known (see [7]) that 0 is stable iff $0 \leq \alpha\rho < \frac{\pi}{2} \approx 1.57$ and asymptotically stable for $0 < \alpha\rho < \frac{\pi}{2}$. Theorem 1 requires $\alpha\rho \leq 1.5$ for stability. Therefore even in the constant coefficients case the constant 1.5 in Theorem 1 can not be much improved. In Theorem 1, the constant $3/2$ cannot be improved (i.e. increased) as the next example shows.

(4) We now show that for $\alpha\rho = 3/2$, (A) can have a periodic solution. Let $a(t,\phi) \equiv 1 = \alpha$ and define $r(t,\phi) = r(t) \leq 3/2 = \rho$ to be periodic of period $5/2$ as follows

$$r(t) = \begin{cases} t & \text{for } t \in [0, 3/2], \\ 3/2 & \text{for } t \in [3/2, 5/2), \\ r(t - 5n/2) & \text{for } t \in [5n/2, 5(n+1)/2) \text{ and } n = 1, 2, \ldots . \end{cases}$$

Then $\dot{x} = -x(t - r(t))$ has a piecewise differentiable periodic solution ϕ with period 5 where

$$\phi(t) = \begin{cases} 1 - t & \text{for } t \in [0, 3/2], \\ 17/8 - 5/2\, t + 1/2\, t^2 & \text{for } t \in [3/2, 5/2], \\ (-1)^n \phi(t - 5n/2) & \text{for } t \in [5n/2, (n+1)5/2) \text{ and } n = 1, 2, \ldots \end{cases}$$

Therefore "$\alpha\rho < 3/2$" in Theorem 1 cannot be improved since in this example $\alpha\rho = 3/2$.

(5) We now see how Theorem 1 can be applied to a complicated non-linear equation. For some $\rho > 0$, let

$$F(\phi) = -\int_0^\rho g(s,\phi(-s))\,d\eta(s)$$

where η is monotonically increasing on $[0,\rho]$ and $\eta(\rho) - \eta(0) = 1$. Let $g : R \times R \to R$ be continuous. Let γ be a continuous monotonically increasing real-valued function such that $\gamma(|x|) > 0$ for $|x| > 0$. Assume for some α, $0 < \alpha\rho < 3/2$ and for some $\beta > 0$

$$|x|\gamma(|x|) \leq xg(s,x) \leq \alpha|x| \quad \text{for} \quad s \geq 0 \quad \text{and} \quad |x| < \beta .$$

CLAIM. 0 is asymptotically stable for

$$\dot{x} = F(t,x_t) = -\int_0^\rho g(s,x(t-s))\,d\eta(s) .$$

Proof. For fixed t and ϕ, let $G(s) = g(s,\phi(-s))$. By the intermediate value theorem in Remark 1,

$$-F(t,\phi) = G(r) = G(-r(t,\phi)) = g(t,\phi(-r(t,\phi)))$$

for some $r = r(t,\phi) \in [0,\rho]$. Define $a(t,\phi) = 0$ if $g(t,\phi(-r)) = 0$ and otherwise let

$$a(t,\phi) = g(t,\phi(-r))[\phi(-r)]^{-1} .$$

Then $\gamma(|\phi(-r)|) \leq a(t,\phi) \leq \alpha$ for all $t \geq 0$ and $\|\phi\| < \beta$. By Theorem 1, all solutions $\psi(\cdot;t_o,\phi)$ with $t_o \geq 0$ and $\|\phi\| < 2\beta/5$ tend to a constant. We now verify (1.3). Choose $|\theta| \in (0,\beta)$ (as in Definition 3).

We consider $\theta > 0$ since the argument for $\theta < 0$ is similar. Choose $\eta_\theta \in (0,\theta/2)$ such that $\eta_\theta + \theta < \beta$ and $\eta_\theta < \gamma(\theta/2)$. Then for $t > 0$ and $|\phi - \phi^\theta| < \eta_\theta$, $\phi(-r(t,\phi)) > \theta/2$ and

$$a(t,\phi) \geqslant \gamma(\phi(-r(t,\phi)) \geqslant \gamma(\theta/2) \geqslant \eta_\theta .$$

Hence a is uniformly non-zero near each non-zero constant, and 0 is asymptotically stable.

(6) The monotonicity result generalizes a result by Kakutani and Markus [9, Theorem 10] which they obtained for a specific equation.

Cooke has also studied in [10] equations with a single lag which depends on x. He has looked for more detailed exponential results as the lag function r tends to 0 in some nice integrable way. The following Corollary follows quite easily from Theorem 1, and it is included here as a more explicit example than is given in Theorem 1. This example emphasizes that Theorem 1 is *not* a linear approximation result since $\frac{d}{dx} g$ need not exist, and even if it does we may have $\frac{d}{dx} g(0) = 0$.

We leave open all questions of *uniform* asymptotic stability.

Corollary 1. *Let* $g : R \to R$ *be continuous,* $xg(x) > 0$ *for* $x \neq 0$, *and let* $r : [0,\infty) \times R \to [0,\infty)$ *be continuous. Then* 0 *is asymptotically stable for*

$$\dot{x}(t) = -g(x(t - r[t,x(t)]))$$

if $r(t,x) \to 0$ *as either* (i) $x \to 0$ *uniformly in* t, *or* (ii) $t \to \infty$ *uniformly for* x *in a compact neighborhood of* 0 .

There are several ways to try to generalize Theorem 1 . One is to consider higher order equations. Alternatively the methods used can be extended to the one-dimensional equation allowing the total lag interval $(-\sup \rho_i, 0]$ to be unbounded. Consider

$$(F_\infty) \qquad \dot{x}(t) = -\sum_{i=1}^{\infty} a_i(t,x_t)\, x(t - r_i(t,x_t))$$

Theorem 2 . *For some* $\beta > 0$, *and sequences (finite or infinite)* $\{q_i\}$, $\{\rho_i\}$, $q_i \geq \rho_i \geq 0$, *let* $a_i : J \times C_\beta^{q_i} \to [0,\alpha_i]$, $r_i : J \times C_\beta^{q_i} \to [0,\rho_i]$. *Assume* $\rho_j \sum_{i=1}^{\infty} \alpha_i \geq 1$ *for all* j *and* $\sum_i^{\infty} \rho_i \alpha_i < 3/2$. *Then* 0 *is stable for* (F_∞) *and for any solution* $\psi(t;t_0,\phi)$ *with* $\|\phi\| < 2\beta/5$, *we have*

$$\psi(t) \to \textit{constant as } t \to \infty .$$

Furthermore if any a_j *is uniformly non-zero near each non-zero constant then* 0 *is asymptotically stable.*

BIBLIOGRAPHY

[1] R. BELLMAN, Research problem: Functional differential equations, Bull. Amer. Math. Soc. 71(1965), 495.

[2] K. COOKE, Functional differential equations close to differential equations, Bull. Amer. Math. Soc. 72(1966), 285-288.

[3] J. HALE and J. LASALLE (editors), Differential equations and dynamical systems, Academic Press, 1967.

[4] J. KATO, Asymptotic behaviors in functional differential equations, Tôhoku Math. Journal, second series, 18(1966), 174-215.

[5] J. KATO, On the existence of a solution approaching zero for functional differential equations, to be published.

[6] T. M. APOSTOL, Mathematical Analysis, Addison Wesley, 1957.

[7] R. BELLMAN and K. COOKE, Differential-difference equations, Academic Press, 1963.

[8] V. I. LOGUNOV, On asymptotic behavior of solutions of the nth order linear differential equation with retarded argument, (Russian), Differencial 'nye Uravnenija 1(1965), 467-468; Math. Reviews, 33(1967) #382.

[9] S. KAKUTANI and L. MARKUS, On the non-linear difference-differential equation $y'(t) = [A - By(t - \tau)]y(t)$, Contributions to the theory of nonlinear oscillations, Vol. 4, edited by S. Lefschetz, Princeton Univ. Press, 1958.

[10] K. COOKE, Asymptotic theory for the delay-differential equation $u'(t) = -au(t - r(u(t)))$, J. Math. Anal. Appl. 19(1967), 160-173.

THE USE OF LIAPUNOV FUNCTIONS FOR GLOBAL EXISTENCE

by

Aaron Strauss

Consider the ordinary differential equation

$$\text{(E)} \qquad x' = f(t,x)$$

where x and f belong to R^n and t is real. For simplicity I shall always assume that f is continuous in (t,x) and locally Lipschitz in x on $J \times R^n$, where J is some interval, possibly unbounded. Thus (E) has uniqueness. If, for $t_o \in J$ and $x_o \in R^n$, the solution $x(\cdot)$ of (E) through (t_o,x_o) exists on $[t_o,\infty) \cap J$, then we say that the solution $x(\cdot)$ *exists in the future*. If we can replace $[t_o,\infty)$ by $(-\infty,t_o]$ above, we say that $x(\cdot)$ *exists in the past*. If we can replace $[t_o,\infty)$ by $(-\infty,\infty)$ above, we say that $x(\cdot)$ *exists forever*.

A function $V : J \times R^n \to [0,\infty)$ is called a Liapunov function if V is continuous in (t,x) and locally Lipschitz in x on $J \times R^n$. If V is a Liapunov function, then we define

$$\dot{V}(t,x) = \limsup_{h\to 0^+} h^{-1}(V(t+h,\ x+hf(t,x)) - V(t,x))$$

$$= \limsup_{h\to 0^+} h^{-1}(V(t+h,\ x(t+h)) - V(t,x)) \quad ,$$

where $x(\cdot)$ is the solution of (E) through (t,x) . More details concerning these derivatives are given in a previous article by Yorke.

The first result that I want to give may be found in [6] as Theorem 3.4, Theorem 3.5, and the remark on page 16. Yoshizawa has informed me that the result was essentially proved by Okamura about 30 years ago.

Theorem 1 . *Let* J *be compact. Then all solutions of* (E) *exist in the future if and only if there exists a Liapunov function* V *such that:*

(1) $\qquad V(t,x) \to \infty$ *as* $|x| \to \infty$ *uniformly for* $t \in J$,

(2) $$\dot{V}(t,x) \leq 0 \ .$$

The restriction that J be compact seems essential in the construction of V used by Yoshizawa. Later, Conti [1] stated a result in one direction which seemingly improved both (1) and (2) . Also, Conti used $J = [0,\infty)$.

Theorem 2 . *Let* $J = [0,\infty)$. *Let* V *be a Liapunov function such that*

(3) $\qquad V(t,x) \to \infty$ *as* $|x| \to \infty$ *for each* $t \in J$

(4) $$\begin{cases} \dot{V}(t,x) \leq w(t,V(t,x)) & \textit{and all maximal} \\ \textit{solutions of} \quad r' = w(t,r) & \textit{exist in the future.} \end{cases}$$

Then all solutions of (E) *exist in the future.*

Unfortunately, Conti's proof was incorrect. His proof does hold for the following result, also proved later by LaSalle and Lefschetz [2, p. 108], which is a special case of Theorem 2.

Theorem 3 . *Let* $J = [0,\infty)$. *Let the Liapunov function* V *satisfy* (4) *and*

$$(5) \qquad \begin{cases} V(t,x) \to \infty \quad \text{as} \quad |x| \to \infty \quad \text{uniformly for } t \\ \text{in every compact subset of } J . \end{cases}$$

Then all solutions of (E) *exist in the future.*

Proof. Suppose the conclusion were false. Then there would exist a solution $x(\cdot)$ of (E) starting at some t_0 and a number $T > t_0$ such that we would have

$$|x(t)| \to \infty \quad \text{as} \quad t \to T^- .$$

Thus, by (5), we would have

$$V(t,x(t)) \to \infty \quad \text{as} \quad t \to T^- .$$

But, from (4)

$$0 \leq V(t,x(t)) \leq r(t)$$

for $t_o \leq t < T$, where $r(\cdot)$ denotes the maximal solution of $r' = w(t,r)$ through $(t_o, V(t_o, x(t_o)))$. Thus we would have

$$r(t) \to \infty \quad \text{as} \quad t \to T^- ,$$

which would contradict (4). Hence the conclusion must be true.

A converse to Theorem 3 was apparently provided by me [4]. This would have been a natural extension of Theorem 1 to $J = [0,\infty)$.

Theorem 4 . *Let* $J = [0,\infty)$. *Then all solutions of* (E) *exist in the future if and only if there exists a Liapunov function* V *satisfying* (2) *and* (5) .

However the proof of the "only if" part of Theorem 4 which I gave is false. Thus two of the first four results in this area have wrong proofs. Stranger yet, there seems to have been an unawareness of previous results. Specifically, Conti says he was not aware of Okamura's result, LaSalle says he was unaware of both Conti's and Okamura's results, and I was unaware of all three.

To motivate the subsequent results that I shall present here, let us note that if $[0,\infty)$ is replaced by $(-\infty,0]$, and if t is replaced by $-t$, Theorem 3 yields

Theorem 5 . *Let* $J = (-\infty,\infty)$. *Let the Liapunov function* V *satisfy* (5) *and*

$$(6)\quad \begin{cases} \rho(t,V(t,x)) \leq \dot V(t,x) \leq w(t,V(t,x)), \text{ where all} \\ \text{maximal solutions of } r' = \rho(t,r) \text{ exist in the} \\ \text{past and all maximal solutions of } r' = w(t,r) \\ \text{exist in the future.} \end{cases}$$

Then all solutions of (E) *exist forever.*

In correcting the proof of Theorem 4, Kato and Strauss [3] were able to prove a converse to Theorem 5.

Theorem 6 . *Let* $J = (-\infty, \infty)$. *Then all solutions of* (E) *exist in the future if and only if there exists a Liapunov function satisfying* (4) *and* (5) . *Furthermore, all solutions of* (E) *exist forever if and only if there exists a Liapunov function* V *satisfying* (5) *and* (6) (*actually with* $\rho \equiv w \equiv 0$).

Finally, the validity of Theorem 2 and the relation between (3) and (5) is established [5] by

Theorem 7 . *Let* $J = (-\infty, \infty)$. *Let* V *satisfy* (3) *and* (4) . *Then all solutions of* (E) *exist in the future. Furthermore, there exists an example showing that* V *can satisfy* (3) *and* (4) *without satisfying* (5) . *Nevertheless, if* V *satisfies* (3) *and* (6) , *so that all solutions exist forever, then* V *must also satisfy* (5) .

This shows that Theorem 2 is stronger than Theorem 3 in spite of Theorem 6. That is, if one wanted to prove that all solutions of (E) exist in the future, then Theorem 6 says there would have to exist V satisfying (4) and (5) . But that particular Liapunov function might be extremely complicated. If one has only been able to produce a Liapunov function satisfying (3) and (4) , then Theorem 3 yields no information while Theorem 2 yields the desired result.

BIBLIOGRAPHY

[1] R. CONTI, Sulla prolungabilità delle soluzioni di un sistema di equazioni differenziali ordinarie, Boll. Un. Mat. Ital. 11(1956), 510-514.

[2] J. P. LASALLE and S. LEFSCHETZ, Stability by Liapunov's Direct Method with Applications, Academic Press, New York, 1961.

[3] J. KATO and A. STRAUSS, On the global existence of solutions and Liapunov functions, Annali Mat. Pura Appl., to appear.

[4] A. STRAUSS, Liapunov functions and L^p solutions of differential equations, Trans. Amer. Math. Soc. 119(1965), 37-50.

[5] A. STRAUSS, A note on a global existence result of R. Conti, Boll. Un. Mat. Ital., to appear.

[6] T. YOSHIZAWA, Stability Theory by Liapunov's Second Method, Math. Soc. Japan, Tokyo, 1966.

A REMARK ON A RESULT OF STRAUSS

by

Junji Kato

1. For a system

$$\dot{x} = A(t)x + g(t,x) , \tag{1}$$

Onuchic [1] has discussed the problem of eventual boundedness and of stability of each bounded solution. Generalizing Onuchic's results, Strauss [2] has considered the same problem for a system

$$\dot{x} = f(t,x) + g(t,x) \tag{2}$$

and obtained the following results: Suppose that $f(t,x)$, $g(t,x)$ are continuous on $[0,\infty) \times R^n$ and $f(t,x)$ has continuous derivatives with respect to x . For the system

$$\dot{x} = f(t,x) \tag{3}$$

we assume that

(i) the system (3) is uniformly stable,

(ii) there exists a bounded solution of (3),

(iii) for any $\alpha > 0$, there exists a constant $M(\alpha)$ such that if $||z|| \leq \alpha$, then

$$\left\|\frac{\partial x(t;z,s)}{\partial z}\right\| \leq M(\alpha)$$

for all t, s, $t \geq s$, where $x(t;z,s)$ is a solution of (3) through z at $t = s$. Moreover, suppose that $g(t,x)$ satisfies the condition

(iv) there exists a continuous function $\lambda(t,\alpha)$ such that if $||x|| \leq \alpha$,

$$||g(t,x)|| \leq \lambda(t,\alpha)$$

and that

$$\int^{\infty} \lambda(t,\alpha)dt < \infty$$

for each $\alpha > 0$.

Then, the solutions of the system (2) are eventually uniform-bounded, and if solutions of (2) are unique for the initial valued problem, each bounded solution of the system (2) is stable.

To obtain this result, Strauss utilized a generalized variation of constants formula due to Alekseev. Here, we shall present some remarks on his results.

2. Under the assumption (iii), we have

(4) $$\|x(t;x_1,t_o) - x(t;x_2,t_o)\| \leqq nM(\alpha)\,\|x_1 - x_2\|$$

for all $t \geqq t_o$, if $\|x_1\|$, $\|x_2\| \leqq \alpha$, and hence (iii) implies (i). Moreover, clearly the assumptions (i) and (ii) imply the uniform boundedness of solutions of the system (3). Therefore since $\frac{\partial x(t;z,s)}{\partial z}$ is a matrix solution of the linear system

$$\dot{u} = A(t,x(t;z,s))u ,$$

where $A(t,x) = \partial f(t,x)/\partial x$, the assumptions (i) through (iii) are equivalent to the assumption

(v) the solutions of the product system

(5) $$\begin{cases} \dot{x} = f(t,x) \\ \dot{u} = A(t,x)u \end{cases}$$

are uniformly bounded.

For example, consider the equation

$$\dot{x} = -x^3 .$$

Then, the corresponding system (5) becomes

(6)
$$\begin{cases} \dot{x} = -x^3 \\ \dot{u} = -3x^2u \ . \end{cases}$$

Since $V(x,u) = x^2 + u^2$ satisfies

$$\dot{V}(x,u) = -2x^2(x^2 + 3u^2) \leqq 0$$

along a solution of (6), the condition (v) is satisfied. For the system

(7)
$$\dot{x} = y - F(x) \ , \quad \dot{y} = -x - g(y)$$

which is obtained from the equation

$$\ddot{x} + f(x)\dot{x} + g(\dot{x} + F(x)) + x = 0 \ , \ F(x) = \int_0^x f(s)ds \ ,$$

we assume that $f(x)$ is continuous, $g(y)$ is continuously differentiable, $f(x) \geqq 0$, $g'(y) \geqq 0$ and $g(0) = 0$. Then, by using the Liapunov function $V(x,y,u,v) = x^2 + y^2 + u^2 + v^2$, we can see the uniform boundedness of the system

$$\begin{cases} \dot{x} = y - F(x), \quad \dot{y} = -x - g(y) \\ \dot{u} = v - f(x)u, \quad \dot{v} = -u - g'(y)v \ , \end{cases}$$

and hence the system (7) satisfies the condition (i) through (iii) .

More generally, since the condition

$$(8)\quad (x - y,\ f(t,x) - f(t,y)) \leqq \ell(t,\alpha)\,\|x - y\|^2, \quad \text{if} \quad \|x\|,\ \|y\| \leqq \alpha ,$$

is equivalent to

$$(u, A(t,x)u) \leq \ell(t,\alpha)\,\|u\|^2 , \quad \text{if} \quad \|x\| \leq \alpha ,$$

where

$$(x,y) = {}^T xy ,\quad (x,x) = \|x\|^2 ,$$

the condition (v) is implied by

(vi) the solutions of the system (3) is uniformly bounded, and there exists a continuous function $\ell(t,\alpha)$ which satisfies the condition (8) and

$$-\infty < \varliminf_{t\to\infty} \int_0^t \ell(s,\alpha)\,ds \leqq \varlimsup_{t\to\infty} \int_0^t \ell(s,\alpha)\,ds < \infty$$

for all $\alpha > 0$. In fact, to prove this it is sufficient to show that solutions of the system

$$\dot{u} = A(t, x(t;x_0,t_0))u ,\quad \|u(t_0)\| \leq 1 ,$$

are bounded by a constant $M(\alpha)$ for any solution $x(t;x_0,t_0)$ of (3) such that $\|x_o\| \leq \alpha$. This can be proved by using a Liapunov function

$$V(t,u) = \exp[-2 \int_o^t \ell(s,\beta(s))ds] \|u\|^2 ,$$

where $\beta(\alpha)$ is a constant such that $\|x_o\| \leq \alpha$ implies $\|x(t;x_o,t_o)\| \leq \beta(\alpha)$ for all $t \geq t_o$.

3. Next, we shall show that it is possible to apply Liapunov's method to Strauss' results. As is shown by the example

$$\dot{x} = \begin{cases} - \sin x \cdot \sin \frac{1}{x} & x \neq 0 \\ 0 & x = 0 , \end{cases}$$

in the converse theorem for the uniform stability of a solution or for the uniform boundedness, the function $V(t,x)$ defined by

(9) $$V(t,x) = \sup\{ \|x(t + \tau;x,t)\| : \tau \geq 0\}$$

for a solution $x(s;x,t)$ of (3) is not necessarily continuous. However, it turns out that under the assumptions (i) and (ii) $V(t,x)$ is continuous, and moreover, under the assumption (iii), or more generally under the condition (4), $V(t,x)$ satisfies a Lipschitz condition. That is, $V(t,x)$ satisfies the following condition

(vii) (a) $\|x\| \leq V(t,x) \leq b(\|x\|)$ for a continuous function $b(r) \geq 0$;

(b) $\dot{V}(t,x) \leq 0$ along a solution of (3) ;

(c) $|V(t,x) - V(t,y)| \leq L(\alpha) \|x - y\|$, for some number $L(\alpha)$ and all $t \geq 0$, if $\|x\|$, $\|y\| \leq \alpha$.

By using this Liapunov function $V(t,x)$, we can conclude that under the assumption (iv) the solutions of the system (2) are eventually uniformly bounded.

Similarly, under the same assumptions the function $W(t,x,y)$ defined by

$$W(t,x,y) = \sup\{ \|x(t + \tau; x,t) - x(t + \tau; y,t)\| : \tau \geq 0\}$$

is continuous and satisfies the condition

(viii) (a) $\|x - y\| \leq W(t,x,y) \leq B(\alpha) \|x - y\|$ for a number $B(\alpha)$ and all $t \geq 0$, if $\|x\|$, $\|y\| \leq \alpha$,

(b) $\dot{W}(t,x,y) \leq 0$ along a solution of

$$\dot{x} = f(t,x)$$
$$\dot{y} = f(t,y) ,$$

(c) $|W(t,x,y) - W(t,x',y')| \leq L(\alpha)\{ \|x - x'\| + \|y - y'\| \}$ for a constant $L(\alpha)$ and all $t \geq 0$, if $\|x\|$, $\|x'\|$, $\|y\|$, $\|y'\| \leq \alpha$.

And the existence of such a Liapunov function guarantees the eventual uniform stability of each bounded solution of the system (2) under the assumption (iv). Hence, if the bounded solution is unique for the initial value problem, then it is uniformly stable.

Conversely, we can show that the existence of both of the Liapunov functions $V(t,x)$ and $W(t,x,y)$ which satisfy the conditions (vii) and (viii), respectively, is equivalent to the condition (v).

Thus, we have the following

Theorem. Let the assumption (iv) *be made. If there exists a Liapunov function* $V(t,x)$ *satisfying the condition* (vii), *then solutions of the system* (2) *are eventually uniformly bounded. Furthermore, in addition, if there exists a Liapunov function* $W(t,x,y)$ *which satisfies the condition* (viii) *and if solutions of the system* (2) *are unique for the initial value problem, then each bounded solution of the system* (2) *is uniformly stable.*

The existence of both of the Liapunov functions above is equivalent to the assumption (i) and the condition (4). If $f(t,x)$ is continuously differentiable, then this is equivalent to the assumption (v). Moreover, the condition (vi) implies the assumption (v).

BIBLIOGRAPHY

[1] N. ONUCHIC, On the uniform stability of a perturbed linear system, J. Math. Anal. Appl. 6(1963), 457-464.

[2] A. STRAUSS, On the stability of a perturbed nonlinear system, Proc. Amer. Math. Soc. 17(1966), 803-807.

SINGLE SPECIES MODEL FOR POPULATION GROWTH DEPENDING ON PAST HISTORY

by

Gregory Dunkel

A mathematical model for population growth was proposed in 1677 by M. Hale and in the 1840's Verhulst wrote a series of papers on the logistics equation;

$$\dot{n}(t) = [b - an(t)]n(t) \ , \tag{1}$$

where n is the density of the population under consideration. Verhulst's model was later extended and modified by Pearl, Lotka and especially by Volterra, who introduced "hereditary effects" or dependency on past history into his formulation.

If we may assume that our growth model can be written as

$$\dot{n}(t) = (b - d)n \ ,$$

where b and d are respectively the birth and death rates of n (not necessarily constant), then the Verhulst model (1) rests on taking

$$b = \text{a constant}$$

$$d = an(t)$$

A simple modification of (1) which allows for "hereditary" effects is

$$\dot{n}(t) = [b - an(t - \gamma)]n(t) \tag{2}$$

Equation (2) has been extensively studied by a series of mathematicians, E. M. Wright, Kakutani and Markus, G. S. Jones, and J. A. Yorke (unpublished). A more complicated, and perhaps more realistic, modification is given by

$$\dot{n}(t) = [b + \int_{\gamma}^{\tau} \psi(n(t - a))dS(a)]n(t) \tag{3}$$

Equation (3) is the subject of our talk.

Biologically, b is a birth rate, assumed constant; $S(a)$ is the "survival factor" or the fraction of n surviving to age (a); τ is the maximum lifespan; γ is the "reaction lag" or the time it takes an increase in population to increase the death rate; ψ measures how <u>much</u> an increase in n increases the death rate.

Mathematically, ψ is some continuous increasing function with $\psi(0) = 0$; $S(a)$ is decreasing, possibly discontinuous and $S(\tau) = 0$; $0 \leq \gamma < \tau < \infty$. Hence, equation (3) includes (1) and (2) as special cases.

We need to make another assumption about equation (3). Notice that equations (1) and (2) have $n = 0$ and $n = b/a$ as critical points. Now $n = 0$ is obviously a critical point of (3) and we assume that a solution n^* of

$$\psi(n) = b/S(\gamma)$$

exists, is unique and positive.

In general, all initial conditions are positive.

Since equation (1) can be integrated explicitly, and easily, it is not too hard to show:

a) $n = 0$ is totally unstable and $n^* = b/a$ is globally asymptotically stable;

b) all solutions are monotonic -- increasing, if $n(0) < b/a$ and decreasing if $n(0) > b/a$.

While equation (2) has the same critical points as (1) , it has a bit more varied behavior:

i) all solutions are uniformly bounded by $\max\{n(0), (b/a)e^{b\gamma}\}$;

ii) for $b\gamma > 1$, all solutions oscillate;

iii) for $b\gamma \leq 3/2$, the solution $n^* = b/a$ is globally stable;

iv) for $b\gamma > \pi/2$, there exists a periodic solution to (2).

(Properties (i) - (iii) are due to Wright; (iv) to Jones.)

It is possible to prove that equation (3) has an analogous set of properties by placing further restrictions on ψ as needed.

We need the following

Definition: A function *oscillates* if it is neither forever monoton. nor a constant.

If a solution to (3) does not oscillate, then using the monotonicity of ψ it is easy to see that it is either decreasing and bounded below by n^* or increasing and bounded above by n^* . If it does oscillate, then it has maxima and using

$$n(t) \leq n^* e^{b(t-z)} \qquad (n(z) = n^*)$$

we see that its maxima are bounded by

$$n^* e^{b\tau} \quad ,$$

and we get

Theorem 1 . *All solutions to* (3) *are uniformly bounded; in fact,*

$$0 < n(t) \leq \max\{n(0) \ ; \ n^* e^{b\tau}\} \quad .$$

Remark. This theorem relates the saturation level, or n^* , the maximum lifespan, τ , and the birth rate of n to an upper bound on its growth.

The next theorem is an oscillation result for solutions of (3) , which depends on the following lemma.

Lemma. *If a solution* $n(t)$ *does not oscillate, then* $n(t) \to n^*$. The discussion before Theorem 1 showed such solutions has a limit and it is not too hard to prove that this limit is n^* .

Now we can indicate how to go about establishing

Theorem 2 . *Suppose* $b\gamma > 1$ *and* $\psi(n^*) = kn^*$. *Then the following implications hold: if* $\psi(n) \leq kn$, *no non-trivial solution is forever increasing; if* $\psi(n) \geq kn$, *no non-trivial solution is forever decreasing.*

Remark. Biologically, $b\gamma$ is the birth rate times the reaction lag and so this theorem could be interpreted as saying whenever there is a sufficiently long lag in a species' reaction and a sufficiently high birth rate oscillations occur. It has been conjectured that in any reasonable, single species growth model, the introduction of a lag leads to oscillation. The problem with settling this conjecture is deciding what makes a model reasonable

Note that $b\gamma > 1$ implies $\gamma > 0$. We need some kind of restriction on γ because for $\gamma = 0$, equation (3) has

$$\dot{n}(t) = [a - bn(t)]n(t) \tag{1}$$

as a special case and *no* solution of (1) oscillates.

Sketch of Proof. Setting $s = t/\gamma$ and considering

$$\dot{n}(s) = [b\gamma + \int_1^{\tau'} \psi(n(s-a))dS'(a)]n(s)$$

we assume $\dot{n}(s) > 0$ for $s \geq T$. Then by the hypothesis that $kn^* = b\gamma/S(\gamma)$ and the lemma, we have

$$\frac{1}{kS(\gamma)} < n(s) < \frac{b\gamma}{kS(\gamma)} \quad ,$$

then

$$\frac{b\gamma}{kS(\gamma)} + \frac{1}{kS(\gamma)} \int_1^{\tau'} \psi(n(s - a))dS'(a) < \dot{n}(s) \ .$$

Using the monotonicity of ψ and $n(s)$ and $\psi(n) \leq kn$, we see

$$\frac{b\gamma}{kS(\gamma)} - n(s - 1) < \dot{n}(s) \ .$$

Integrating this differential inequality from $T + 2$ to $T + 3$ gives

$$\frac{b\gamma}{kS(\gamma)} < n(T + 3) \ ,$$

a contradiction.

Corollary. Assume the same hypotheses, except that

$$\psi(n) \begin{cases} \leq kn & (n \leq n^*) \\ \geq kn & (n \geq n^*) \end{cases}$$

then all solutions oscillate.

We can also get a result on the global stability of n^* .

Theorem 3 . *Assume* $\psi(n^*) = kn^*$. If

$$\psi(n) \begin{cases} \geq kn & (n \leq n^*) \\ \leq kn & (n > n^*) \end{cases}$$

and $b\tau \leq 1$, *then* $n(t) \to n^*$ *as* $t \to \infty$.

The proof of Theorem 3 is quite messy and tricky, so we won't attempt to sketch it. Comparing the last two theorems, notice that oscillation required conditions on the birth rate and reaction lag, while stability requires conditions on the birth rate and lifespan. Futhermore, the conditions on ψ in Theorem (3) are just the obverse of those on ψ in Theorem 2.

It is also possible to show that equation (3) has periodic solutions.

Theorem 4 . *If* $\gamma > 0$ *and* b *is sufficiently large, and if* ψ *satisfies*

$$\psi(n) \begin{cases} \leq kn & (n \leq n^*) \\ \geq kn & (n > n^*) , \end{cases}$$

then equation (3) *has a periodic solution.*

The proof of this theorem depends on a close analysis of the zeros of $n(t;\phi) - n^*$, when $n(t;\phi)$ is a solution of (3) whose initial condition is in the class S_k of all ϕ in $C^1[-\tau,0]$ with the properties:

a) $0 < \phi(s) < n^* e^{ks}$, $-\tau \leq s < 0$ and $\phi(0) = n^*$

b) $|\dot{\phi}(s)| \leq M$, where M is independent of ϕ .

If we can show that the variable translation operator

$$T\phi[s] = n(s + z_3 \,;\, \phi)$$

(z_3 is the third zero of $n(t;\phi) - n^*$) is continuous and an into map, that is,

$$T : S_k \to S_k ,$$

then we can appeal to Schauder's Fixed Point Theorem and the uniqueness of solutions in order to pick out a periodic solution, which is the fixed point of T .

This method of attack is due to G. S. Jones.

AN EXTENSION OF CHETAEV'S INSTABILITY THEOREM USING INVARIANT SETS AND AN EXAMPLE

by

James A. Yorke

In this lecture I am presenting an improved version of Chetaev's Instability Theorem. (See for example [4, Theorem 3.6.23]). My version of Chetaev's Theorem is concerned only with the autonomous equation

$$\text{(A)} \qquad \dot{x} = f(x)$$

where $f : U \to R^n$ is continuous, U is open in R^n, $0 \in U$ and $f(0) = 0$ In some cases where f is defined on a "large" set, it is worth while to restrict f to a small open neighborhood of 0, while letting U be some ε-neighborhood of 0. The application in the Example $(\ddot{y} = F(y,\dot{y}))$ probably uses the simplest Liapunov function ever used $(V(y_0,y_1) = y_0)$ for a non-trivial result. A special case of this example was given by Leighton [6]. His assumptions on $\frac{\partial}{\partial \dot{y}} F(0,0)$ exclude the unstable equation $\ddot{y} = y$, and his conclusions are weaker. These results were discovered while I was looking for applications of invariant sets. Several other applications can be found in [1]. The Example is an extension of Theorem 4.3 in [2]. I make no attempt to generalize Theorem 2 or the Example to non-autonomous equations. Any such attempt should probably take into consideration the non-autonomous generalizations of LaSalle's Theorem on Liapunov functions, first by generalized Yoshizawa and then by LaSalle [5, p. 277-286].

Notation and Assumptions. For simplicity the solutions of (A) will be assumed unique, and $\phi(\cdot,x)$ will denote the solution such that $\phi(0,x) = x$. A set G is called *positively (negatively) invariant* if for each $x \in G$ and $t \in$ domain $\phi(\cdot,x)$, $t > 0$, $(t < 0)$, we have $\phi(t,x) \in G$. (If uniqueness were not assumed, the invariant and positively invariant sets would have to be changed to *weakly* invariant and *weakly* positively invariant in Theorem 2 and Remark 2. See [1] for the definitions of weak invariance.) 0 is *unstable* if for some $\varepsilon > 0$ and every $\delta > 0$ there exists x, $|x| < \delta$, such that for some $t \geq 0$ in the domain of $\phi(\cdot,x)$, $|\phi(t,x)| > \varepsilon$.

The set $G \subset U$ is assumed closed relative to U, i.e., $\bar{G} \cap U = G$, and "∂G" denotes the boundary of G in U, so $\partial G \in U$, and $V : G \to [0,\infty)$ is assumed locally Lipschitz with $V(0) = 0$. As usual $\dot{V}$ denotes the trajectory derivative

$$\dot{V}(x) = \liminf_{\tau \to 0+} \tau^{-1}[V(x + \tau f(x)) - V(x)]$$

(Of course, the function $\overset{*}{V}$, introduced in my first lecture, p. 35, could be used if we only had V continuous. The results would still hold). I first state the usual Chetaev Theorem for autonomous equations.

Theorem 1. *Assume*

(1.1) $V \equiv 0$ *on* $\partial G \cap U$ *and* $V > 0$ *on* $G - \partial G$;

(1.2) $\dot{V}(x) > 0$ *when* $V(x) > 0$;

(1.3) $\bar{G}$ *is the closure of an open set and* $0 \in \partial G$.

Then 0 *is unstable.*

Theorem 2 . *Let* $G_v = \{x \in G : V(x) = v$ *and* $\dot{V}(x) = 0\}$ *and* $G_+ = \{x \in G : V(x) > 0\}$. *Assume*

(2.1) G *is positively invariant*;

(2.2) $0 \in \bar{G}_+$, $V(0) = 0$, *and* $\dot{V} \geq 0$ *on* G;

(2.3) *for* $v > 0$, G_v *has no (non-empty) compact invariant subsets.*

Then 0 *is unstable.*

Furthermore if

(2.4) G_0 (i.e., G_v *with* $v = 0$) *has no compact invariant subsets except* $\{0\}$,

then

(2.5) *there exists a solution* $\phi \not\equiv 0$ *defined on* $[0,\infty)$ *such that* $\phi(t) \to 0$ *as* $t \to -\infty$.

Whenever the assumptions of Theorem 1 are satisfied, those of Theorem 2 are also satisfied, and it will frequently be easier to find V if Theorem 2 is used. The purpose of the condition (1.1) in the usual proof of Theorem 1 is in part to guarantee that G is positively invariant. I believe it is wiser to directly assume as in (2.1) that G is positively invariant and so discard a strong restriction on V . Positive invariance for a specific set G is usually easy to verify from knowing only G and f . We don't have to know what the solutions are. Indeed necessary and sufficient conditions on G and f for positive invariance when solutions are unique are given in a theorem by M. Nagumo. (See [1, Theorem 2.1]). In [3, Theorem 5(b)], Massera's (non-autonomous) version of Chetaev's Theorem closely approaches my assumption that G is positively invariant by using an analytic condition which is harder to verify and is only a special case of positive invariance; however, his theorem contains the *idea* of making G a positively invariant tx set. Unlike Chetaev's and Massera's, my assumptions do not assume G contains an open set, and so Theorem 2 can be applied, for example, even when V is defined only on a lower-dimensional positively invariant set containing 0 .

The second innovation in Theorem 2 copies LaSalle's theorem for stability by assuming $\dot{V} \geq 0$ and that there are no troublesome invariant sets where $\dot{V} = 0$, rather than assuming (1.2). The third innovation is more original, namely using (1.4) to get the additional result (1.5). It is also the only part of the theorem that is at all difficult to prove. (*Any* difficulties in proof should make us happy since the "direct Liapunov theorems" tend to be a theory of the "nearly obvious.") The Example makes use of all three changes and is quite difficult to handle without these changes. I know of no V which could be used to show 0 is unstable for (S) or (S') using Theorem 1.

Example. Let $F : U \to R$ be continuous and solutions of (S) be unique, where $U \subset R^2$ is a neighborhood of (0,0) . Consider

(S) $$\ddot{y} = F(y,\dot{y})$$

Assume $F(0,0) = 0$ and $F(y,0) > 0$ for $y > 0$. Then there is a solution $y(\cdot) \not\equiv 0$ such that $y(t) \to 0$ and $\dot{y}(t) \to 0$ as $t \to -\infty$, and $(0,0)$ in the $y\dot{y}$ plane is unstable.

Proof. The following system is equivalent to (S):

$$\text{(S')} \qquad \begin{aligned} \dot{y}_o &= y_1 \\ \dot{y}_1 &= F(y_o,y_1) \; . \end{aligned}$$

Let $V(y_o,y_1) = y_o$. Then $\dot{V}(y_o,y_1) = y_1$.
Let $G = \{(y_o,y_1) : y_o \geq 0$ and $y_1 \geq 0\}$. G is positively invariant for (S') since for $y_1 = 0$, $y_2 \neq 0$, or for $y_1 \neq 0$, $y_2 = 0$, the vector $(y_1, F(y_o,y_1))$ points into the interior of G . Clearly if a soluti $\phi(0) \neq 0$ but $\phi(0) \in \partial G$, then $\phi(t)$ is in the interior of G for small $t > 0$; hence no solutions in U can leave G . Note that G_+ contains the interior of G so $0 \in \bar{G}_+$, and $G_v = \{(v,0)\}$ is clearly not invariant for $v > 0$ since $F(v,0) > 0$. Hence 0 is unstable by Theorem 2. To see that there is a solution $\phi = (y(\cdot), \dot{y}(\cdot))$ of (S') as claimed note that $G_o - \{0\}$ is empty so (2.4) is satisfied.

Remarks. (1) Consider the system $\ddot{y} = 0$, or $\dot{y}_o = y_1$, $\dot{y}_1 = 0$. Let $V(y_o,y_1) = y_o y_1$, and $G = \{(y_o,y_1) : y_o \geq 0 , \; y_1 \geq 0\}$. Then (2.1), (2.2), (2.3) are satisfied, as are in fact (1.1) and (1.2), but (2.4) and (2.5) both fail. Hence, (2.5) is not a consequence of the usual Chetaev theorem.

(2) By changing signs in Theorem 2 we can conclude the following:

If G is *negatively* invariant, $0 \in \bar{G}_+$ and $\dot{V}(x) \leq 0$ on G ; if for all $v \geq 0$ G_v has no compact invariant subsets except perhaps $\{0\}$, then there is a solution $\phi \not\equiv 0$ such that $\phi(t) \to 0$ as $t \to +\infty$.

Now by letting $G = \{(y_o,y_1) : y_o \geq 0 , y_1 \leq 0\}$ in the Example and again $V(y_o,y_1) = y_o$ we find there exists a non-zero solution ϕ , $\phi(t) = (y(t) , \dot{y}(t))$, such that $\phi(t) \to 0$ as $t \to +\infty$.

(3) In the Example, it suffices to have $F(y,0) > 0$ for $y \in (0,\varepsilon)$ for some $\varepsilon > 0$. Then we can choose $U = \{(y_o,y_1) : |y_o| < \varepsilon\}$. By examining the third quadrant instead of the first and letting $V(y_o,y_1) = -y_o$, it suffices to have $f(y,0) < 0$ for $y \in (-\varepsilon,0)$ for some $\varepsilon > 0$ for the conclusions in the Example to hold.

BIBLIOGRAPHY

[1] J. A. YORKE, Invariance for ordinary differential equations, Math. Systems Theory 1(1967), 353-372.

[2] J. A. YORKE, Asymptotic properties of solutions using the second derivative of a Liapunov function, Univ. of Maryland Ph.D. dissertation, August 1966.

[3] JOSÉ MASSERA, Contributions to stability theory, Annals of Math., 64(1956),182-206.

[4] N. BHATIA and G.P. SZEGÖ,Dynamical Systems: Stability Theory and Applications, Lecture Notes in Mathematics #35, Springer-Verlag, 1967.

[5] J. HALE and J. P. LASALLE, Differential Equations and Dynamical Systems, Academic Press, 1967.

[6] WALTER LEIGHTON, On the construction of Liapunov functions for certain autonomous nonlinear differential equations, Cont. Differential Equations, 2(1963), 367-383.

Offsetdruck: Julius Beltz, Weinheim/Bergstr.